ENCYCLOPAEDIA OF BACTERIAL DISEASES IN PLANTS

ENCYCLOPAEDIA OF BACTERIAL DISEASES IN PLANTS

Vol. 3

Shahzad Ahmad

ANMOL PUBLICATIONS PVT. LTD.
NEW DELHI - 110 002 (INDIA)

ANMOL PUBLICATIONS PVT. LTD.
H.O.: 4374/4B, Ansari Road, Darya Ganj,
New Delhi-110 002 (India)
Ph.: 23278000, 23261597
B.O.: No. 1015, Ist Main Road, BSK IIIrd Stage
IIIrd Phase, IIIrd Block,
Bangalore - 560 085 (India)
Visit us at: www.anmolpublications.com

Encyclopaedia of Bacterial Diseases in Plants

First Published, 2008

ISBN 978-81-261-3532-5 (Set)

PRINTED IN INDIA

Printed at Mehra Offset Press, Delhi.

Contents

Preface

Bacterial diseases are generated by bacteria. These are very small organisms, not seen by naked eyes. The bacteria are disseminated in infected seeds and travel plant to plant and field to field through blowing winds splashing rains, sprinkler irrigation, surface-drainage water, insects, birds, animals and human beings, etc.

The life period of bacteria may be 6 to 18 months on or above the soil surface under dry conditions. There are two types of bacterial infections—primary infection and secondary infection. Primary infection involves the invasion of seedlings as they emerge from the soil. Secondary infection is due to the spread from a primary source to other growing plants. Bacteria enter a plant through natural leaf openings, and through wounds created by hail, blowing soil particles, insects or cultivator injury.

The diseases typically develop, when contaminated seeds are planted, when plantings are made in fields with a history of the diseases and when the climate is consistently hot and wet or humid. Pod lesions start as round, water-soaked dots that enlarge, merge, dry, and form sunken, irregular, frequently reddish brown blotches. When severe, entire pods may be badly shrivelled and die. Seeds in less affected pods develop normally and show no signs of diseases. When seeds containing bacteria are germinated, many of them fail to germinate. Those, which germinate produce seedlings with blight lesions on the cotyledons, stems, and first true-leaves. The stems of the seedlings may have water-soaked, suken areas that enlarge and develop into reddish streaks. Any time during the season affected stems commonly crack and become griddled by water-soaked cankers or rots.

Diseased plants may be dwarfed, with the upper leaves crinkled and mottled. Dark green, water-soaked areas may also be visible on the stems. Young seedlings, infected with halo blight develop a chlorosis, which may be general or confined to the young trifoliate leaves at the top of the plant. The lesions may merge and their centres fall out giving leaves a tattered appearance. When bacterium becomes systemic, tan sunken lesions with a reddish brown margin formed on the stems and petioles. Water-soaking is not as pronounced as it is in common and halo blights. Later, the grey centres may crack open or drop out, leaving small holes. The veins on the under leaf surface turn red or reddish brown. The spots on the stems and pods are somewhat more elongated then those on the leaves.

This research-based work has a perfect description of bacterial diseases in plants in an interesting manner. Expectedly, this work would be able to make a separate stream in the study of bacterial diseases in plants.

Editor

1

Growth of Antibiotics

Early Experiments

Those growth inhibitors or bactericidal agents that are formed by living organisms, as opposed to synthetic chemicals or salts, and that are active in very low concentrations, are called *antibiotics.* It is recognised that no clear line can be drawn between the synthetic and natural groups; phenol is, in the last analysis, a biological product (being obtained from coal tar), while chloramphenicol is at the moment being synthesised on a large scale. Thus, the term *antibiotic* cannot be too rigidly defined.

"Antagonism" of one microorganism by another has been noticed from early times, and Bryson (1962) records that as early as 1885 Cornil and Babes wrote: "If the study of the mutual antagonism of bacteria were sufficiently far advanced a disease caused by one bacterium could probably be treated by another." E. Duchesne, in a Lyon thesis of 1897, found that a Penicillium reduced the virulence of *E. coli* and *Salmonella* when injected into an animal with them, and concluded: "In pursuing the study and facts of biological antagonism between moulds and bacteria one may arrive perhaps at other facts applicable to prophylactic hygiene and to therapy." Several instances of microbial antagonism were studied before the recent great development of antibiotics began. One arose from the discovery that the damping-off fungus, *Rhizoctonia solani,* which kills many seedlings, is itself killed by a common soil organism, *Gliocladium fimbriatum.* This led to the isolation of Gliotoxin (Weindling and Emerson, 1936; Butcher *et al.*, 1945), the first natural antibiotic to be prepared pure. Some of the other cases involved deliberate efforts to develop antagonistic organisms or antibiotics, using masses of bacteria as substrate. From such cultures Dubos (1939) isolated a sporeformer, *Bac. brevis,* which inhibited the growth of pneumococci by means of a soluble excretion product. Higher concentrations of the substance actually lysed the cocci. From the culture medium two bactericidal substances, gramicidin and tyrocidin, were isolated. These early experiments are well reviewed by Waksman (1943) and by Florey *et al.*, (1949).

Penicillin

The first really effective antibiotic was nearly overlooked. In 1928, Fleming, in studying variation in Staphylococcus "noticed that around a large colony of a contaminating mould the staphylococcus colonies became transparent and were obviously undergoing lysis" (Fleming, 1929). The organism was later identified as *Penicillium notatum* West ling and was found to inhibit streptococci also. Extracts were prepared, and the substance named penicillin. Yet, although Fleming called attention to its potential value "in the treatment of pyogenic infections," medical men showed no interest.

It was only when Florey and co-workers at Oxford showed that infected mice could be protected from death by injections of the crude penicillin that clinical trials began to be made. Finally, the war needs stimulated an international project to isolate and purify the active material (Raper, 1952, or Fleming, 1956).

Many Penicillia other than the original organism produce penicillin. One of the early high-yielding strains belonged to another species, *Pen. chrysogcnum,* and besides many other members of this variable group, 12 other species of Penicillium and seven of Aspergillus produce penicillin. Penicillin is also formed by the almost unrelated Trichophyton and by *Malbranchea puJchella*. The strains used for commercial production represent selected mutants, produced by x-ray and ultraviolet treatment from a naturally occurring, high-yielding variant. Many laboratories cooperated in irradiating the cultures and screening the mutants for high-yielding strains.

The Penicillia are grown in liquid culture in a current of sterile air in an acid glucose medium. After a few days of vigorous growth, when the glucose is exhausted, there is added another carbohydrate, such as lactose, which is less rapidly assimilated, and the medium made less acid. This results in very slow further growth and rapid excretion of penicillin.

A complication is that some organisms produce the enzyme penicillinase, which inactivatives penicillin by partial hydrolysis to penicilloic acid. This makes it necessary to harvest the cultures at an optimal time and calls for strict control of the process throughout. Alkali causes the same hydrolysis.

Chemical purification of the antibiotic (Florey *et al.*, 1949) led, after many difficulties, to the establishment of structure for penicillin, where R may represent aryl or alkyl groups. The most commonly occurring R groups are benzyl and p-hydroxybenzyl, pentenyl, amyl, and heptyl. The resulting different penicillins. All are active, but different organisms or media produce different mixtures.

The related *Cephalosporin* C, produced by a species of the imperfect fungus *Cephalosporium,* has a 6-membered thiolactam ring, and the R group is a-aminoadipyl (Abraham and Newton, 1956). When *a*- aminoadipic acid is added to the medium, it is incorporated into the R group; cystine similarly goes into the ring (Trown *et al.*, 1963). After the structures were elucidated, and it became clear that penicillin G (R = benzyl) was of particular value as an antibiotic, Moyer suggested that an attempt be made to increase the yield of this compound by providing the fungus with phenylacetic acid. It was indeed found that the total yield of penicillin could thus be increased, and also that cultures producing a high proportion of penicillin K could be made to yield largely penicillin G instead. In current practice, phenylethylamine or phenylacetic acid esters are used (Tabenkin *et al.*, 1952). Many other "biosynthetic" penicillins have been made in the same way; e.g., adipic acid yields carboxy-rc-butyl penicillin, which has about the same activity against gram-negative forms but much less than penicillin G against gram-positives.

Other constituents of the media that favour high penicillin yields are less closely related to its structure; amino acids tend to increase the yield, but cysteine, which is clearly part of the molecule, has little effect (Cook and Brown, 1950). Plant extracts such as corn-steep liquor or pea extract are mainly used for medium, and synthetic penicillin is not economically practical.

In the last few years, many new penicillins have been made by hydrolysing off the benzoyl group from ordinary penicillin G and treating with acyl chlorides to introduce new R groups (Batchelor *et al.*, 1959). The hydrolysis of the amide linkage uses *penicillin amidase* from several organisms, including *Str. lavendulae,* and was first found to occur spontaneously in penicillin manufacture. Some of the resulting "semi-synthetic" penicillins are fairly resistant to penicillinase (Novick, 1962), and others are absorbed more rapidly than any of the natural penicillins. Some other more complex derivatives are resistant to acid hydrolysis (Doyle *et al.*, 1961).

A development of the work on penicillin was the discovery in *Pen. notatum* of an enzyme oxidising glucose, called *notatin* or glucose oxidase (Coulthard *et al.*, 1945). It is also formed by *Pen. resticulosum* and other species and has been crystallised from *Pen. amagasakiensis* (Kusai *et al.*, 1960). This enzyme is a flavoprotein containing FAD, which is reduced by the substrate and reoxidised by oxygen; the glucose is converted to gluconic acid:

> The accumulation of H_2O_2 might inhibit the growth of the Penicillium, as it does in the case of lactic bacteria; however, since the Penicillia contain catalase, the H_2O_2 is either decomposed or used to oxidise alcohol. Because of this enzyme, *Pen. notatum*, which produces alcohol from sugars, forms during its growth a mixture of alcohol, acetaldehyde, and gluconic acid.

Another related discovery is the enzyme penicillinase, which is formed adaptively, especially by *Bac. cereus*. It occurs in three forms, one extracellular, one internal but readily washed out, and one tightly bound (Pollock, 1956; Sheinin, 1959). In many organisms this enzyme is the basis for penicillin resistance. Fortunately, it can be assayed with great sensitivity (Novick, 1962).

Its action is to open the 4-membered ring. Penicillin is remarkably non-toxic to man and other mammals, the jio (i.e., the dose killing 50 per cent of the test animals) being as high as 1 to 2 gm per kilogram body weight. As an antibiotic, it is especially effective against cocci, both gram-positive arid gram-negative, including pneumococci, streptococci, meningococci, and gonococci; it is also very effective against the clostridia of gas gangrene and against *Bac. anthracis*. Most important from a social viewpoint is its effectiveness against the Treponemas of syphilis and yaws. Thus, both the major venereal diseases (gonorrhoea and syphilis) are brought under control for the first time. In the South Pacific, penicillin has brought down the incidence of yaws from some 15 per cent of the population to under 1 per cent by 1960.

Penicillin acts mainly against organisms that are actively growing and not merely surviving. This fact has been put to use for isolating nutritional mutants (Davis, 1949), which fail to form some metabolite and therefore, require it in the medium.

If a culture containing a few such mutants is grown on minimal medium, all the unchanged cells will start to grow, and if penicillin is now added, the growing cells will be killed and only the mutants will survive. (The dosage of penicillin must be carefully graded.) Penicillinase can then be added to destroy the penicillin, and the residual cells will then grow.

Methods of Assay

Chemical Methods: When an antibiotic is available in pure form, and its structure is at least partially known, the amount of it can be determined by chemical methods based on its reactive groups. For instance, the lactam ring of penicillin, I, can be hydrolysed with alkali to sodium penicilloate, II, and the amount of base consumed can then be determined by backtitrating. Or the penicilloic acid formed on hydrolysis can be broken down further with acid to yield dimethylcysteine, III, which can be determined polarographically or by the ninhydrin reaction for amino acids. There are other possibilities also (Lester Smith, 1948).

For streptomycin similar methods are possible, e.g., the very reactive CO group can be converted to a fluorescent hydrazone (Boxer and Jelinek, 1947). Tetracyclines can be determined by their ultraviolet absorption (Hiscox, 1951) or better by the fluorescence of their complexes with calcium and barbiturate (Kohn, 1961).

The most specific methods, however, are those based on antibiotic power. These, of course, can be used with impure preparations at every stage from the first crude extracts onward. Three types of procedure are possible, as follows.

All-or-none Growth Method

Tubes of nutrient broth are seeded with a standard small inoculum of a susceptible organism and serial dilutions of the antibiotic added. The tubes are incubated and examined. The lowest concentration necessary to inhibit growth completely is taken as end point, and the amount of antibiotic determined by comparison with similar dilutions of a known standard substance. For penicillin, *M. aureus,* incubated 16 hours at 37° C, is used. An advantage of this method is that the volumes used can be quite small.

Turbidimetric Method

Here again tubes of broth are seeded with a small inoculum, but the time course of growth is followed with a turbidimeter or colorimeter. In this curve, using penicillin, the higher concentrations cause autolysis after four hours, as shown by decreasing density. A plot of the turbidity or optical density f log y J at a fixed time against concentration is a curve; the same plotted against log.

Diffusion Methods

These methods are the ones most particularly associated with antibiotics. A plate of solid medium, inoculated in bulk with test organisms, is treated in spots with the solution and then incubated. As the substance diffuses out into the agar, it inhibits growth throughout a zone surrounding the place of application. The radius of the zone is a measure of the amount or concentration of the antibiotic. If the antibiotic applied is small in amount, the radius of the inhibited zone is proportional to its *amount*. This is true of a small drop of solution, and also, roughly, with small filter-paper circles, soaked in the solution. However, if a relatively large volume of solution is used, the radius of the zone is proportional to the *concentration,* or more nearly to the logarithm of the concentration, of the antibiotic in it.

This is the case with the classical "Oxford cup" method, in which small glass cylinders, their ends warmed in the flame, are placed on the agar and filled with 1 to 2 ml of the test liquid. When this method is used with penicillin, *Bac. subtilis* is used, since bulk inoculation of agar has to be done at 48°C, a temperature that may damage a coccus (Foster and Woodruff, 1944-1946).

The original Oxford unit of penicillin was established by the cup assay. It was defined as that amount of penicillin which, under the standard conditions, using *"Staph." aureus,* gave an inhibition zone of diameter 24 mm.

Streptomycin and Congeners

A systematic study of the antibiotics produced by the *Streptomycetaceae,* begun by Waksman and co-workers at Rutgers University about the beginning of World War II, has led to the discovery of many valuable antibiotics. Over 400 of them have been isolated, and about 30 of these have found use in medicine. In addition, Bryson (1962) lists 70 more, from Aburamycin to Virocidin, all but one formed by Streptomyces species, and mostly discovered since 1957. Waksman (1952) noted that about 20 per cent of all Actinomycetes isolated from soil had *some* depressing effect on the growth of other microorganisms.

The first two compounds obtained, *actinomycin,* from *Str. antibioticus,* and *streptothricin,* from *Str. lavendulae,* were bactericidal in vitro but were too toxic to be useful in medicine; however, *actinomycin* is now being tried for inhibiting the growth of tumours. The third, *streptomycin,* obtained from *Str. griseus,* was extremely powerful and was relatively non-toxic (Schatz *et al.,* 1944). It is a characteristic of penicillin that it affects principally the gram-positive organisms, but streptomycin is active against both gram-negative and gram-positive bacteria, which gives it wider clinical application.

In particular, it strongly inhibits *Mycob. tuberculosis,* which is unaffected by penicillin. Many tuberculosis patients have been more or less cured by streptomycin, but some forms of the disease are resistant, and tuberculosis still ranks with malaria as the world's major infectious disease. Among other human diseases susceptible to streptomycin is tularemia, caused by an organism resembling the Pasteurella of plague. This disease is completely controlled by streptomycin if administered soon enough after infection (Chapman *et al.,* 1949).

With monkeys, treatment beginning 48 hours after a heavy infection caused complete recovery, while controls all died in four to six days. Diseases caused by the coli-aerogenes bacteria, including one type of bacterial pneumonia *("Klebsiella")*, are also susceptible to streptomycin.

A serious difficulty with streptomycin treatment, which holds to some extent for all antibiotics, is that continued treatment with the drug may lead to the appearance of streptomycin-resistant strains, by selection. Such selection may take place very quickly in culture media containing the antibiotic, since the difference in growth rate between the sensitive culture and the resistant mutant is so very great.

Some of the resistant mutants encountered among pathogens are just as virulent as the original culture and grow at about the same rate. For this reason they become established as a permanent part of the population (Davis, 1952), with the result that diseases treated by antibiotics tend to become steadily less amenable to the treatment. The ultimate in the development of resistance is selection of a strain that actually requires streptomycin for growth.

To some extent, the development of resistant strains of *Mycob. tuberculosis* is prevented by administering p-aminosalicylic acid, or isonicotinic hydrazide, between the streptomycin injections, for these inhibit growth also (Long and Ferrebee, 1950), and the chance that the double mutation will appear, giving resistance to both compounds, is slight. From its first production on a small scale in 1946, the manufacture of streptomycin has increased very rapidly. Thousands of tons a month are now produced in the United States, and large factories operate in England, Japan, Denmark, Germany, etc., as well. Steel tanks, aerated with sterile air, are used as vessels, and the medium is usually a soybean extract with the addition of dextrose, $CaCO_3$, and a vitamin concentrate from cereal grains, such as "distiller's solubles" or cornsteep liquor (Ledingham, 1953).

The yield of the original *Str. griseus* culture was found to decrease with time, but mutant strains have been produced by irradiating with ultraviolet and X-rays, and by treating with "nitrogen mustard." However, the yields have been increased in this way only by a factor of about eight (Dulaney, 1953). A serious difficulty in streptomycin production is the development of bacteriophage ("actinophage") attacking *Str. griseus*. Phage-resistant strains have been developed, but, unfortunately, these usually produce lower yields of antibiotic. The chemistry of streptomycin has been completely elucidated (Brink and Folkers, 1950). It contains three moieties:

> *Streptidine*, an inositol derivative with two guanidine substituents *Streptose*, a methyl pentose of unusual branched structure *Streptosamine*, which is N-methyl-L-2-glucosamine. These three structures are linked by ether bonds.

Such curious complexity led to the comment (Kluyver, 1952) that: "It can most aptly be described as the product of an organic chemist just before his admittance to a lunatic asylum. The structures of each of the three moieties of the molecule are equally irrational and the mere idea that nature has thought fit to combine these elements to form a molecule with remarkably specific action tends to depress the most modest among biologists." While it is true that the structure is highly unusual, still, it does suggest some possible bases for growth inhibition.

First, the two guanidine groups recall the powerful anti-malarial activity of diguanidine compounds like Paludrine, and indeed when both these groups are hydrolysed off by alkali, the product is antibiotically inactive. When one is removed, about 10 per cent of the activity remains (Schaffner, 1961). Second, the inositol structure in Streptidine reminds us that many organisms require inositol for growth. Inhibition by streptomycin is even partly reversed by adding Streptidine (Hancock, 1960). Third, Streptosamine resembles the L- isomer of the glucosamine and N-acetyl glucosamine in cell walls, and there is some evidence that streptomycin acts on the cell wall, although this is still uncertain.

As to the biosynthesis of streptomycin, the Streptosamine portion is formed from glucose, since C^{14} in the 1 and 6 positions in glucose added to the medium is found in high yield in the same positions in the Streptosamine (Silver-man and Rieder, 1960). How the D-configuration is completely inverted to the L- remains unknown. The inositol ring in Streptidine is derived from myoinositol, which enhances production of the antibiotic and decreases the incorporation of C^{14}-glucose into the Streptidine (Majumdar and Kutzner, 1962). The guanidine groups apparently come from CO_2, which suggests that they are formed from arginine (Hunter *et al.* 1954), and added arginine does sometimes increase the yield. The streptose is probably derived from rhamnose, which is found in *Sir. griseus* extracts as a nucleotide, thymidinediphos- phate-rhamnose (Baddiley *et al.*, 1961).

Streptomycin may also occur in three modifications. One is a mannoside, called streptomycin B; this can be hydrolysed by a mannosidase. Another, hydroxystreptomycin, is produced by *Str. griseocarneus;* its general properties are similar to those of streptomycin. Third, much of the streptomycin used in medicine is in the form of the dihydro derivative, in which the aldehyde group of streptose is reduced; this has somewhat less toxicity, but its activity, at least in tuberculosis, is undiminished (Brink and Folkers, 1950). Its activity in vitro against *M. aureus* is the same as that of streptomycin and of the mannoside (Hancock, 1960).

Related to streptomycin is the mixture of three substances formed by *Str. fradiae* 3535 and called *neomycin*. One of these has the formula like streptomycin, it is active against both gram-positive and gram-negative bacteria (Waksman and Lechevalier, 1962), but it has not been used clinically because of toxicity to the kidneys. It is formed in much larger amounts than streptomycin, yields of up to 8 gm per litre being obtained industrially on a rich medium (Upjohn, 1958). Kanamycin is similar (Umezawa *et al.*, 1958).

Streptothricin also occurs usually in a mixture, being conjugated with various numbers of molecules of the non-protein-forming amino acid, beta-lysine. Streptothricin itself contains D-gulosamine and the base, streptolidine, making it of similar structure to streptomycin. The Actinomycins are peptides and will be taken up below.

Chloromycetin or Chloramphenicol

This antibiotic is produced by an organism obtained originally from Venezuelan soil (Smith *et al.*, 1948), and hence called *Streptomyces venezuelae.* The composition is relatively simple, and is remarkable for containing a nitro group and two Cl atoms, both very rare in natural products (Bartz, 1948). Indeed, only one other natural nitro compound, nitropropionic acid, is known; large yields of this have been isolated from *Pen. atrovenetum* (Raistrick and Stossl, 1958), and it occurs in higher plants too.

The two asymmetric carbon atoms of chloramphenicol give rise to four optically active isomers, of which only the D-isomer is highly active. A racemic mixture is readily synthesised, so that the chloramphenicol now used is being made synthetically. It is remarkably stable, resisting nine hours' boiling in neutral solution (Gillissen, 1950) and hence very useful in surgical work.

Many gram-negative organisms are sensitive to chloramphenicol, including the Enterobacleriaceae. It is also used against gonococci and spirochetes, where these are resistant to penicillin. A special value is its action against rickettsiae, including those causing Rocky Mountain spotted fever, scrub typhus, Q fever, and typhus. However, it has recently come into disfavour for clinical use, due to a very slowly appearing anaemia.

Tetracyclines

The great effectiveness of streptomycin led, in the late 1940's and 1950's, to a systematic search of soils from all over the world for new species of Streptomyces with antibiotic properties. Several hundred growth-inhibiting compounds have been purified from the resulting cultures; many had only very weak activity, and many turned out to be identical with streptomycin or other known antibiotics. Of the few that are really effective the tetracyclines are outstanding. Chlorletracycline or Aureomycin, isolated from *Sir. aureojaciens* (Duggar *et al.*, 1948), and oxytetracycline or Terramycin from *Str. rimosus* (Sobin *et al.*, 1950) are chemically very similar.

Both are produced on well-aerated media, of which corn or soy beans are the main constituent. Both are highly active against gram-negative rods and of very low toxicity to animals. Aureomycin particularly is very rapidly absorbed, circulated, and excreted, and hence is fully effective when taken orally; gastrointestinal, urinary, and peritoneal infections are readily susceptible to it. Both act to some extent against the "microorganisms," including those of virus pneumonia, influenza, and some Rickettsiae. Terramycin is usually accompanied by a second antibiotic, rimocidin, which inhibits many fungi. The tetracyclines are so closely related that the same organism often produces both.

With NH_4Cl in the medium, *Str. aureofaciens* makes a mixture that is 90 to 95 per cent chlortetracycline, but if *pH* compounds are added, the chlorination is inhibited and tetracycline results (Goodman *et al.*, 1959).

Peptide Antibiotics

Many sporeformers excrete polypeptides which are powerful antibiotics. The most active organisms are *Bac. brevis, Bac. licheniformis, Bac. subtilis,* and *Bac. polymyxa*. Some streptomycetes also form active peptides, especially *Str. antibioticus, Str. parvullus,* and *Str. chrysomallus.* These produce the Actinomycins.

Tyrothricin is the mixture of antibiotics produced by *Bac. brevis,* which was the organism that developed when pneumococcus cells were used as an enrichment substrate

(Dubos, 1939). It contains mainly *tyrocidin* and *gramicidin*, both of which have been crystallised (Hotchkiss, 1944). Gramicidin varies in composition; a Russian preparation, gramicidin S, contains only 10 amino acids, two each of valine, ornithine, leucine, D-phenylalanine, and proline, linked in a closed cycle. Other gramicidins contain a few more amino acids in the ring, including D-leucine.

Both gramicidin and tyrocidin cause kidney damage and are too toxic for internal use but are often applied to bandages and dressings. *Bacitracin* is an unfortunate name, derived from "*Bac. subtilis* Tracy" (Johnson *et al.*, 1945). This organism has also been called *Bac. licheniformis* and is undoubtedly related to the true *Bac. subtilis* Cohn, although distinguished from it by the ability to grow anaerobically in the presence of nitrate.

Other strains of *Bac. subtilis* produce *subtilin* (Jansen and Hirschmann, 1944), *bacillin* (Foster and Woodruff, 1946), and *mycobacillin* (Majumdar and Bose, 1960). All these are mainly active against gram-positive bacteria and fungi.

Several other preparations differ in detail (Lockhart and Abraham, 1956). In general, they are too toxic for internal use. However, bacitracin is sometimes used on infections resistant to other antibiotics where the treatment can be closely observed and is also used superficially for "sterilisation" of the skin and of the intestines prior to surgery (Raper, 1952). Since bacitracin has the same effect on farm animals as the other antibiotics, it is being used in fair quantities as a feed supplement.

Polymyxin and *Aerosporin* are polypeptides produced by *Bac. polymyxa*, another name for which is *Bac. aerosporus.* These comprise a group of at least five polypeptides, each produced singly by a different strain. *Circulin*, produced by *Bac. circulans*, is very closely related. The polymyxins are highly active against gram-negative bacteria, but again toxicity to the kidney has so far limited their use in internal medicine. They have been used, however, in diseases of farm animals and in human patients with infections resistant to streptomycin. The chemical composition of polymyxin B (Biserte and Dautrevaux, 1957) is that of a cyclic octapeptide with a long side-chain, (di-ab = *Colimycin)* produced by *Bac. colistinus,* is another cyclic decapeptide with five molecules of diaminobutyric acid in the ring, and a similar side-chain. Polymyxin A contains D-leucine as well, while polymyxin D has both D-leucine and serine. All are highly bacteric rather than bacterios. Because of the diaminobutyric acid, they are all strongly basic, as are tyrocidin, gramicidin, and subtilin also.

This basic character makes their action comparable with that of basic detergents and dyes; when the free amino groups are substituted or acylated, the bactericidal activity disappears (Newton, 1958). Not only the basic groups but also the cyclic structure is important, since a straight-chain peptide with the same amino-acid sequence as gramicidin S has less than one-tenth its activity (Erlanger *et al.* 1954). The *Actinomycins* are a group of yellow and red compounds containing the tricyclic phenoxazine ring plus two cyclic peptides:

> Six others are known in which the amino acids between threonine and sarcosine are different (Brockmann, 1960; Johnson, 1960). Since the amino acid, sarcosine, is not a protein constituent, it is missing from most media; by adding it, the yield of actinomycins can be increased (Katz and Goss, 1959).

Similarly, L-valine also increases the yield, but its isomer, D-valine, decreases it (Katz *et al.*, 1961). These compounds are produced by a group of 15 or more streptomycetes whose taxonomy and characteristics are rather confused (Woodruff and Waksman, 1960). They act 100 to 1000 times more strongly on gram-positive than on gram-negative bacteria, but are clinically unusable because of their great toxicity.

Now, however, they are of renewed interest because they are more toxic to tumours than to normal tissue (Hackmann, 1952-1960). As yet, the actinomycins are the only antibiotics used to treat cancer.

The role of cyclic peptides is evidently very wide, since penicillin itself is in reality a cyclic dipeptide, comprising valine and serine, with the C_r-C_7 side-chain. The cyclic anhydride of serine amide OCH_2-CHNHo (Cycloserine or oxamycin) has very similar properties NH -CO.

(Ciak and Hahn, 1959). It seems almost certain that all these compounds act by interfering with the laying down of the cellwall peptide.

Secondary Effects of Antibiotic Feeding

The powerful action of tetracyclines against gram-negative rods, coupled with their great ease of absorption and excretion, often results in almost complete sterilisation of the intestine. In man, bacteria sometimes almost disappear from the faeces. Other antibiotics have similar, though less striking, effects. Occasionally, the sterilised faeces become infected with the yeast *Candida albicans*, which is pathogenic and causes serious secondary infections.

In farm animals, on the other hand, the secondary effects of feeding antibiotics are of great value. It was the feeding of vitamin concentrates that led to this discovery. Since some Actinomycetes produce vitamin B_{12} in large amounts, vitamin concentrates were made from the antibiotic fermentation wastes and .fed to cattle, but it was soon noticed that the resulting growth promotion was greater than that due to the vitamin B_{12} alone. The amount of tetracycline needed for maximum growth promotion is about 10 gm per ton of feed, for chicks, turkeys, and pigs (Raper, 1952). This is about 1000 times the BIZ requirement, which is only 4 to 8 mg per ton. The young animals grow more rapidly and are freer from disease. Other antibiotics, especially penicillin and bacitracin, have similar effects. All are being used in farm feeds. The great penetrability of chlortetracycline has the odd result that when fed to hens, it can increase the resistance of the eggs to Rickettsiae.

It seems hardly likely that the action results from the inhibition of intestinal bacteria, since penicillin is not very active against these, and because the effective concentrations are too low. More probably the antibiotics act by preventing low-level infections (e.g., by Clostridia) which are ordinarily scarcely detected, although they decrease growth.

A danger of this whole development is that animals fed antibiotics may develop highly resistant strains of intestinal bacteria. This has been proven in one case, where the strain of *E. coli* from turkeys fed on streptomycin was 1000 times as resistant to streptomycin as is normal (Starr and Reynolds, 1951). The spread of antibioticresistant pathogens is obviously undesirable. The use of antibiotics in medicine has raised similar problems in the appearance of resistant strains of human pathogens, antibiotic-resistant cocci being now particularly troublesome in hospitals (Verwey, 1959).

Miscellany

With so many effective compounds stopping the growth of bacteria, it is not surprising that many treated patients contract infections with fungi. This is particularly the case in the intestines where normally bacteria reign supreme. As a result, pathogenic fungi have recently become a serious pest, and a search has been made for anti-fungal substances. These should eventually have great agricultural use against fungal diseases of crops.

The most effective anti-fungals so far found are a group of unsaturated large molecules called polyene macrolides. They contain amino sugars and p-aminophenyl residues. *Nystatin,* with 46 C atoms and four double bonds, inhibits growth of many fungi and some algae and prevents glycolysis by yeast (Sutton *et al.,* 1961). The group with seven double bonds, including *candidin* ($C_{46}H_{7}5NO_{17}$, Vining and Taber, 1956), *candicidin, perimycin, trichomycin* and *amphotericin,* is more potent, particularly on *Candida albicans,* one of the principal pathogens. The growth of this fungus is inhibited at 0.5 mg antibiotic per litre, or less (trichomycin at 0.03 mg per litre), while most bacteria are scarcely affected at all (Hosoya *et al.,* 1952).

Because the polyenes are rather insoluble in water, they have been made soluble by N-acetylation, but this lowers the activity by 5 to 10 times (Lechevalier *et al.* 1961). These compounds may prove very useful. Three other water-insoluble types are of interest. The first is the *Oligomycin* group, formed by *Str. diastatochromogenes* (Masamine *et al.*, 1958). Three of these have formulas $C_{24}H_{40}O_6$, $C_{22}H_{36}O_6$, and $C_28H_{40}O_6$, and are neutral unsaturated alcohols. They are active against several skin fungi, as well as the fungi causing oak wilt and Dutch elm disease.

Second is the peculiar compound *Elaiomycin,* from *Str. hepaticus,* which is effective only against mammalian *Mycob. tuberculosis.*

Third is another nitrogenous compound, *Micrococcin,* $C_{25}H_{28}0_6N_6S3$, soluble neither in water nor in alcohol (Heatley *et al.*, 1951, 1952).

It can be given intravenously as fine particles which persist in the body for months. Although it inhibits many bacteria, including *Mycob. tuberculosis,* it is ineffective against experimental tuberculosis.

Netropsin, from *Str. netropsis,* is a different type of compound, since it appears to be a purine derivative and thus may be comparable with the synthetic purine antagonists in (Waller *et al.*, 1957). Although the growth inhibition it causes is not reversed by adding purines, it can be completely reversed in *E. coli* by adding a mixture of 18 amino acids (Zygmunt, 1961); thus it may interfere with the amination process. Chemical aspects of many other antibiotics are treated by Woodward (1957), and biological aspects by Bryson (1962).

The actions of the principal antibiotics on different organisms are presented as a series of overlapping areas.

Development of Resistance

When a bacterium is grown in a series of subcultures in the presence of gradually increasing concentrations of antibiotic, the most resistant cells among the population are selected, and a resistant strain is built up. These strains commonly have characteristically changed morphology, or staining, or become highly filamentous, as though growth can take place but cell division is inhibited. Resistance of bacilli to Actinomycin is accompanied by loss of the ability to sporulate. The phenomenon of "drug resistance" is not limited to the antibiotics but is often exhibited against other growth inhibitors.

The adaptation of *E. coli* to various substances through successive transfers. It is evident that there is little or no adaptation to $HgCl_2$ phenol, or KCN, marked adaptation to the dyes, and very rapid adaptation to streptomycin; in other words, the organism can adapt to some substances but not to others. As a very similar example, *Aerobacter aerogenes,* after 100 subcultures in phenol, did not develop the slightest resistance to phenol, whereas a very few transfers made it resistant to proflavine or to sulphonamide (Hinshelwood, 1949). Resistant strains may become dependent on the drug, especially with streptomycin. Development of resistance to one antibiotic is not independent of resistance to another.

With *M. aureus,* resistance to almost any antibiotic confers some resistance to streptomycin, but not to penicillin or tetracycline; resistance to tetracycline confers resistance to subtilin (McVeigh and Hobdy, 1952). Strains of *E. coli* resistant to strep-tothricin and nëomycin are resistant to streptomycin, but not vice versa, while strains resistant to polymyxin and circulin are resistant to neomycin too (Szybalski and Bryson, 1952). Some strains very highly resistant to chloramphenicol are resistant to tetracyclines (Brock, 1961).

It is important to note that the development of resistance does not require the antibiotic (or other growth-inhibiting agent) to be present. The only effect of the antibiotic is as selecting agent. The situation is exactly comparable to that of the phage-resistant mutants. This important conclusion has been established in several ways, perhaps most neatly by "replica plating" (Lederberg and Lederberg, 1952). An agar plate bearing a number of colonies is pressed with a piece of fine velveteen; the "pile" acts like several thousand inoculating needles, and when pressed on a new agar plate containing antibiotic, it transfers to the plate separate inocula from each of the colonies.

Of these, only the resistant colonies grow. Thus, transfers can be made from all the individual colonies in the parent plate without ever exposing them to the antibiotic, and it is clear that some of these colonies are already resistant. That indeed they *all* are, i.e., that truly adaptive drug resistance is excluded, is not proven by this method, but was later shown by testing a large number of small samples from liquid cultures (Cavalli-Sforza and Lederberg, 1956). Using streptomycin and chloramphenicol, this more laborious method demonstrated that *all* the resistant mutants had arisen before exposure to the drugs.

Only with penicillin is resistance truly inducible. The degree to which drug resistance is developed depends both on the drug and on the organism; for streptomycin the gram-negative rods (with the exception of *Sh. dysenteriae)* develop much greater drug resistance than most of the gram-positive organisms. From *E. coli* strain K_{12}, mutants have been isolated, by periodic transfer to higher concentrations, that resist chloramphenicol at 1000 mg per litre (Brock, 1961).

Resistance to antibiotics can be transferred from one strain to another (1) by sexuality, as in *E. coli,* (2) by transduction, as in Salmonella and *M. aureus,* and (3) by transforming factors. Such resistance is one of the most convenient genetic "markers" to work with. In one bacterium, *M. aureus,* resistance to penicillin, to streptomycin, to chlortetracycline, and to novobiocin can all be transduced (Ritz and Baldwin, 1958; Morse, 1959; Pattee and Baldwin, 1961).

In the special case of penicillin, resistance can be due to hydrolysis of the drug by penicillinase. The transduced resistance is of this type. Formation of penicillinase is readily induced by brief exposure to penicillin or indeed to some of the "semi-synthetic" penicillins (Novick, 1962); it is then excreted into the medium, probably by some enzymatic system in or near the cell wall (Pollock, 1961). Streptomycin resistance in a few cases was traced to a decreased uptake; this occurs also in absence of air (Hancock, 1960).

Mode of Action

The gross site of action of penicillin (but not its exact mechanism) is now understood; it *prevents formation of the cell wall*. When rodshaped cells are placed in penicillin broth, they become spheres and then lyse (Bonifas, 1954); if the broth contains 0.3 M sucrose plus penicillin, they are slowly converted to stable spheres (Lederberg, 1956-1957). Dividing cells swell first at the place where the septum is forming; this delicate crosswall is thus the most susceptible point. (For the same reason some organisms fail to form cross-walls in penicillin broth and grow into long filaments).

When the sucrose broth containing the spheres is diluted, they burst at once. The spheres are evidently *protoplasts* like those produced by lysozyme, which have their membrane intact but no wall. However, in *E. coli* and *Aerob. cloacae,* the spheres can gradually revert to normal cells, so that it is believed that some primordium of the wall remains, and they are termed *spheroplasts* (Hugo and Russell, 1960- 1961). As a rule, they cannot divide, but they can be infected by phages, which grow in them normally.

Earlier studies (Gale, 1949) showed that penicillin-resistant strains of *M. aureus* have lost the ability to accumulate amino acids from the medium. The most resistant strains also become gram-negative, which indicates a change in wall structure. Park and his co-workers (1952, 1957) also found that a few minutes' exposure to penicillin causes *M. aureus* to excrete complex bases, containing uridine and acetyl amino sugars. Evidently the amino sugar was that which *could not be laid down as wall* and hence was excreted (in combined form) instead.

The wall mucopeptide must therefore, be the site of penicillin action. More direct proof is given by an isolated and washed cell-wall preparation from *M. aureus*, which has been shown to incorporate C^{14} glycine, alanine, lysine, or glutamate, the four main amino acids of the wall mucopeptide (Mandelstam and Rogers, 1959). This incorporation is drastically inhibited by penicillin. Since penicillin does not affect protein synthesis by these cells, its action is evidently quite specifically exerted on the wall substance. Molecular models strongly suggest that penicillin binds to the wall at the same site as the N-acetyl-muramic acid (Collins and Richmond, 1962). The N atom common to the two rings of penicillin falls on the O atom of the propionic sidechain in muramic acid, the two carboxyls (which have the same dissociation constants) exactly coincide, and the C:O of the amide bond in penicillin falls on the C:O of the acetyl group of muramic acid. Thus penicillin may well prevent the insertion of acetylmuramic acid into the wall.

It is no wonder that its action on gram-negative cells is weaker, because their walls have a very different composition, *n-cycloserine* has similar action, converting *E. coli* to spheroplasts and causing uridine nucleotides to accumulate. One of these differs from the compound formed in presence of penicillin in not containing two D-alanine residues, and added D-alanine prevents accumulation of this and other nucleotides. Strominger (1962), therefore, concludes that D-cycloserine competes with D-alanine and prevents it from being laid down in the wall peptide.

Chloramphenicol has a quite different action. If *M. aurcus* is fed C^{14}-amino acids in presence of chloramphenicol they are taken up but not incorporated into protein (Gale and Folkes, 1953). Similarly in *£. coli,* protein synthesis is stopped completely by 10 mg of the drug per litre of medium (Wisseman *el al.* 1954). It thus stops synthesis of all enzymes, even including the galactosidase formed in cell-free preparations (Novelli *et al.*, 1961). Chloramphenicol has no effect on the incorporation of glycine into the adenine and guanine of nucleic acids, and indeed the RNA content, at least of the ribosomes, may even be increased, because they contain less protein (Dagley and

Sykes, 1960). Neither is the attachment of amino acids to soluble RNA much affected; it may also be increased.

Chloramphenicol must therefore, inhibit the *final* polymerisation step in protein synthesis (Lacks and Gros, 1960), i.e., the transfer of aminoacyl RNA to the ribosomes. As a result of this, the particles come to contain less and less protein, and hence the RNA becomes less and less stable (Dagley *el al.*, 1962). Apparently most of the ribosomal RNA is normally stabilised by protein, and if not, then its nucleotides are excreted. *Puromycin* has a similar effect, interfering with the transfer from soluble RNA to protein (Yarmolinsky and de la Haba, 1959). In cell-free preparations from *E. coli,* it clearly prevents the condensation of amino acids to proteins on the ribosomes (Nathans and Lip- mann, 1961) and appears to cause the release from the ribosome surface of protein material that is being synthesised (Morris *et al.* 1962).

In Ps. fluorescens, where it also stops protein synthesis, it does not affect the formation of the RNA. The role of the *tetracyclines* is unclear, but they bind strongly to metal cations and thus to the metals bound by proteins or DNA (Kohn, 1961). In bacilli, they inhibit sporulation, preventing uptake of Ca and hence formation of calcium dipicolinate (Brock, 1962; Vinter, 1962). The metal-bound tetracycline is probably held firmly at some crucial site on the cell membrane.

The action of the peptide antibiotics has been foreshadowed by the similarity between the basic dyes and polymyxin. Briefly, polymyxin E causes *Ps. fluorescens* cells to leak pentose and phosphate out into the medium; it also allows a dye, ordinarily not taken up, to enter and combine with protein, producing a fluorescence which occurs only together with protein (Newton, 1953-1954). Polyvalent cations, which decrease permeability, protect the cells against polymyxin. Gramicidin inhibits the uptake of phosphate by M. *aureus* which also suggests action at the membrane (Hotchkiss, 1944). These basic peptides thus probably attach to the membrane and so prevent the incorporation of its normal constituents, making it too permeable as a result.

The polyenes, which act mainly as anti-fungals, have very similar action, since many of them strongly inhibit phosphate uptake by *Candida albicans* and other fungi (Henis and Grossowicz, 1960). Nystatin causes K^+ to leak out from yeast cells too (Sutton *et al.*, 1961). Quite different is the action of *Mitomycin C*, produced by *Str. caespitosus;* this drug was found by Japanese workers (e.g., Mizutani *et al.*, 1959) to inhibit the synthesis of DNA. Curiously, infection with bacteriophage reinstates the DNA synthesis, but the resulting phage particles are not infective (Sekiguchi and Tagaki, 1960), so perhaps their DNA has been modified. A transforming DNA of *Bac. subtilis* was 90 per cent inactivated in about an hour (Nakata *et al.*, 1961). The basis for the action seems to lie in the stimulation of DNA hydrolysis in the cell, since the drug does not itself affect DNA or DNA-ase in vitro.

In contrast to these rather well-defined actions, the role of *streptomycin* remains unclear. Too many effects seem to have been observed:

1. Streptomycin-resistant strains of *E. coli* and *M. aureus* oxidise pyruvate only to acetate and seem unable to condense it with oxalacetate to form citrate (Umbreit, 1949, 1953; Oginsky, 1953). Similar accumulation of acetate occurs in more sensitive strains in presence of the drug. Other strains require pyruvate or acetate, although hematin will substitute, being perhaps used to form an oxidising enzyme system (Jensen, 1953). The rates of oxidation of succinate, lactate, pyruvate, and glucose are all decreased, in both gram-positive and gram-negative forms (Jackson, 1958; Hancock, 1961). All these facts might mean that an oxidising step, or perhaps the "condensing enzyme", is the target of streptomycin's attack.
2. A quite different result is that uptake of amino acids by Salmonella is prevented, and cells growing on an

ammonium medium are resistant to the drug (Orskov, 1960).

3. Incorporation of C^{14}-tyrosine into protein is prevented in extracts of a Mycobacterium, while resistant strains of this bacterium showed no such effect (Erdos and Ullmann, 1959-1960).
4. Resistant strains contain more RNA than sensitive strains in *M. aureus,* 100 per cent more, and in *E. coli* some 20 per cent more (Beljanski, 1952). Strains that *require* streptomycin also have increased RNA content, which streptomycin lowers to the normal level (Spotts and Stanier, 1961). These last three phenomena suggest interference with some stage of protein formation, probably the reaction between amino acids and RNA; however, the fifth set of data points to an entirely different locus of action.
5. Labelled streptomycin is rapidly adsorbed by both sensitive and resistant strains of *E. coli,* reaching completion in one minute, both at 0°C and at 37°0C (Anand *et al.,* 1960). From this physical fixation, which almost certainly occurs on the wall, the streptomycin can be eluted by buffer solutions, i.e., it is held in "ionic linkages," probably by its strongly basic guanidine groups and the phosphate groups in the wall. But after 10 minutes the sensitive cells begin a further major uptake which is slow and prevented by chloramphenicol. Apparently now the streptomycin held on the wall or membrane has damaged its permeability control, and this allows the secondary entry of more streptomycin. Addition of toluene (known to damage cell membranes) causes a similar secondary uptake. Thus, the action would be exerted by opening the cell membrane, perhaps like the action of polymyxin and other basic peptides. Correspondingly, streptomycin is found to cause the excretion of nucleotides and of amino acids. How all these

observations can be unified is not clear. It is remarkable how easily resistance to streptomycin appears in cells cultured in its presence, and still more remarkable how often such resistant types become dependent on streptomycin for normal growth. Such dependence suggests that the cell so readily modifies itself, to accommodate the change caused by streptomycin, that the modification is at least as good as the original form. Spotts and Stanier (1961) suggested that the streptomycin that penetrates into the cell combines with the ribosomes,. making them unable to react with the messenger- RNA and therefore, unable to form proteins. (Probably it would substitute for a normal ribosome constituent). However, a slight modification in the ribosomes can reduce their affinity for the drug, making them resistant. But this also makes them less able to react with the RNA. Now the structure can be further modified to accept the streptomycin as part of it; thus, the ribosome can function again, and the cells are now *dependent* on streptomycin. So the drug must have considerable similarity to the structure of the ribosome surface, and this, it may be suggested, gives it similarity to the membrane structure too. Along some such lines the multiple effects observed may be explained.

The discovery and development of antibiotics have brought a revolution in medical practice. In the 1880's the discovery and culture of the principal pathogenic bacteria led to a revolution in the idea of disease, but the advantage to the sick patient came only slowly, through asepsis, public health measures, and the development of immunology. In the period since 1930, the antibiotic revolution has brought much more immediate results, since it has given us, not the cause, but the cure.

Unfortunately, the antibiotics have not taught us much about the natural interactions between microbes. The simple facts that *Pen. notatum,* producing penicillin, is less common than *Pen. italicum* which forms no antibiotic, while *Str. griseus* was unknown until recently, suggest that antibiotics play little part in natural selection, since otherwise the antibiotic-producing organisms would long ago have become dominant in a competitive world. Indeed, Waksman (1961) points out that Streptomycetes live in the soil, where antibiotics are probably not produced, and where they would in any event be quite unstable.

The antibiotics have taught us something about research, however. The vitamin analogues, dyes, sulphonamide, etc., were developments from what was already known about enzymes and growth factors; they represented an application and perfection of basic knowledge already in hand. The antibiotics, by contrast, were quite new compounds acting in unknown ways. Penicillin and streptomycin are of such curious structures that they would virtually never have been found by routine testing of synthetic compounds. Correspondingly, they have proven very much more valuable in medicine than the previously known inhibitors, and in microbiology they are helping to make comprehensible many and varied aspects of the life of bacteria. Truly, the persistent following up of a new biological lead "opens unknown doors" much more effectively than the refinement of methods and ideas already in existence.

The Evolution of Bacteria of the most fascinating aspects of the study of any group of organisms is its evolutionary history. In the vertebrates and in the higher plants, evolution has been largely deduced from the fossil forms, and by comparison of the details of anatomical structure. The first of these methods is unavailable for bacteria. It is true that bog iron ore shows spiral structures ascribed to the action of ancient Gallionellae or related iron-depositing bacteria, but this is too isolated a case to be very useful, and, besides, the iron oxide is an excretion and not exactly a part of a fossil bacterium. As to anatomy, there is beginning to be assembled a good deal of knowledge about the bacterial cell and its structure, but much of this is at the chemical level and in any event its value for comparative study is still quite limited. The few anatomical points which help to place the bacteria in regard to other organisms are:

(1) The small size of the cell;

(2) The mode of division, i.e., by transverse fission;

(3) The simple structure of the nucleus, with no visible membrane;

(4) The non-cellulosic cell wall, with acetyl-glucosamine prominent; and

(5) The flagella.

Features (1), (2), and (3) emphasise relationship with the blue-green algae and (4) with the fungi. So many unicellular organisms have flagella that the interpretation of (5) is doubtful. The flagellum of an alga like Euglena is so thick and complex, consisting of many strands associated in a kind of rope, that its similarity to those of the bacteria is uncertain (Houwink, 1951).

The absence of flagella in either the Myxobacteria or the blue-green algae suggests that the gulf separating these from the Eubacteria may be greater than appears. The same conclusion can be drawn from the absence of the rigid cell wall in the Myxobacteria and Spirochetes.

Evolution of Life in General

Our interpretation of the evolution of bacteria must depend largely on how we visualise the evolution of organisms in general. Huxley regarded the chemo-autotrophic bacteria as primitive because they produced their kind from inorganic substrates only. Yet they require a complex outfit of enzymes to do this, and most of them, being aerobes, require oxygen-consuming enzymes as well.

The more recent view of evolution, adumbrated by Benjamin Moore in 1912 and worked out in detail by Oparin (1957), is that organic matter was first formed by purely chemical reactions, promoted by the high-intensity radiation which was then reaching the earth. The needed concentration of materials at local spots may have taken place, as Bernal (1951) has suggested, by adsorption at the surface of colloidal particles. The simple organic materials thus produced would have been relatively stable because of the absence of microorganisms to attack them, and so would have gradually given rise to molecules of greater complexity.

Thus, it was only over long periods of time that the first proteins, and thence the first enzymes, were gradually synthesised. It follows that the first organism would have been a heterotroph, metabolising perhaps very slowly the organic compounds that surrounded it. Furthermore, there is some reason to believe that, as the earth cooled down, the large bulk of the oxygen became combined as metallic oxides, carbonates, and sulphates, so that the atmosphere contained little or no oxygen. Our present 20 per cent of oxygen, on this view, arose largely as a product of photosynthesis, although loss of H_2 to the highest atmosphere layers (Miller and Urey, 1959) also played a part. Thus, the first organisms were probably anaerobes.

The reality of these ideas has been greatly strengthened by experiments in which an electric discharge is passed through a mixture of CO_2, NH_3, and H_2, and the gases circulated through water; several organic acids and amino acids, especially glycine and *a*- and \$-alanine, and some guanidine derivatives were produced (Miller, 1953-1957; Miller and Urey, 1959; Heyns *et al.*, 1959).

Using X-rays, the number of products is enriched by the occurrence of direct carboxylation, because CO_2 combines with ethanol to form lactic acid, with methanol to form glycollic acid, and with acetic acid to form malonic acid (Scholes *et al.*, 1960). These reactions, which involve free radicals, are prevented by O_2 and therefore, could only have occurred in the primeval anaerobiosis.

CO_2 solutions alone can produce formaldehyde and formic

acid when subjected to high intensity UV or Y-rays (Getoff, 1962). The simultaneous presence of glycine and organic acids is important because it could have readily given rise to *porphyrins,* since the initial condensation to porphyrinogen occurs spontaneously and *anaerobically.* Hence the porphyrin nucleus of the photosynthetic pigments may well have arisen very early in chemical evolution. How optical activity entered into all the products remains a mystery.

If we accept this broad picture, some interesting consequences ensue. With limited amounts of organic matter, selection would enormously favour organisms capable of synthesising complex substances from simple ones, and especially would favour any kind of photosynthesis, however primitive. Photosynthesis probably came when the supply of organic matter was at its lowest ebb, and it will have first replenished, then enormously increased the supply.

The photosynthetic bacteria are anaerobic; the green algae and higher forms are aerobic. However, some algae, although normally aerobic, can carry out photosynthesis anaerobically ("photoreduction") using hydrogen. This suggests that the development of the chlorophylls resulted in a sequence of organisms first using photosynthesis primarily to increase their supply of synthetic energy, then to convert CO_2 to organic matter, and only later to dispense with oxygen acceptors and thus to produce free oxygen. The characteristic of aerobes is to possess the cytochrome system, which consists of hemin derivatives chemically related to chlorophyll.

Since cytochromes play a part in all photosynthesis, including that of the anaerobes, and are present too in the anaerobe Desulfovibrio, they have probably undergone a *parallel evolution* in their electron acceptors, first functioning with chlorophylls and sulphate and only much later with oxygen itself. Thus, the colourless aerobes, able to produce energy by consuming oxygen, could have appeared on earth only after photosynthesis had produced quantities of *both* organic matter *and* oxygen for them to live on (DeLey in Symposium, 1962).

The path of oxidation of hexose sugars may sometimes lead directly from hexose via phosphogluconic acid, but in very many organisms it leads from pyruvic acid; in other words, it leads from a product of fermentation. This supports the concept that aerobiosis is a later development, superimposed on fermentation.

Evolution and Classification of Bacteria

With this as a working outline, the evolution of the bacteria can be considered. The first true organism contained, within some kind of membrane, (1) enzymes to ferment organic compounds, (2) apparatus (presumably the high-energy phosphate system) to transfer some of the energy to synthetic reactions and to the uptake of more organic matter, and (3) systems to channel the syntheses not only to organic matter in general but specifically to the formation of more enzymes. This last constitutes a problem, since present-day organisms need nucleic acids to synthesise protein; yet proteins (enzymes) are required to form nucleic acids. Perhaps, in the beginning, enzymes were synthesised directly from amino acids by proteinases. The nucleic acids could then have come only later, and still later that "division of labour" (Lanham, 1952) whereby the DNA achieved the trick of self-duplication, while the RNA's became effective as the machine for protein synthesis (Hoagland, 1960). The development of even the simplest true organism (as defined today) would have been very slow.

The first would have been a colourless anaerobe, doubtless of the simplest morphology, and from it would have arisen the following improvements:

(1) Motility, enabling a wider survey of food supplies;

(2) Spore formation, enabling survival through drought;

(3) Gradual increase of synthetic powers, enabling the use of simpler substrates;

(4) Varied morphology, apparently in itself of little evolutionary value, but leading to larger size, and hence to multicellular types;

(5) Photosynthesis, a major step forward, increasing the amount of organic matter on the earth, and eventually giving rise, as a second photosynthetic line (Stanier and van Niel, 1941) to cells producing oxygen in photosynthesis, i.e., the blue-green and higher algae; and

(6) Aerobiosis, including chemosynthetic autotrophy as well as heterotrophy of various kinds.

Except that (5) should precede (6), these developments need not have occurred in the order shown.

The proposed increase in synthetic powers (3) would of course occur by single steps (Horowitz, 1945). Exhaustion of a particular organic compound in the primordial substrate would selectively favour appearance of a mutant able to synthesise it from a simpler precursor. This process endlessly repeated would lead step by step to the development of the present chains of single-enzyme reactions, as exemplified by the arginine or histidine series controlled by closely linked genes.

The final step in this increase in synthetic powers is the synthesis of organic matter from inorganic, but this requires the introduction of a major energy source, supplied either by oxidation or by light.

On the other hand, the development of *control* systems like feedback inhibition and enzyme repression must have come later, through the evolutionary pressure caused by competition. Cells with such control over the formation of their chemical contents and their enzymes obviously have an evolutionary advantage because they can develop faster or more economically on a given substrate than cells without such control. Besides, they maintain a balance between those constituents like the amino acids (or purines) that are needed in fixed relative amounts to synthesise specific polymers.

The requirement of "simplest morphology" would make the primitive form a coccus. This agrees with the fact that very few cocci have either spores or flagella. The anaerobic cocci, like Veillonella and Ruminococcus with their propionic-type fermentations, or Methanococcus with its great ability to reduce CO_2, would then be close to the primitive type, although very likely not themselves of primitive origin. Lactic streptococci probably do not belong here but with the lactic rods.

It is probable that the rigid cell wall of the Eubacteria is not a primitive character. If the primitive organism had no rigid cell wall, it might well have given rise to two lines, one becoming the Eubacteria and the other the Myxobacteria. However, the Myxobacteria are all rods in the vegetative form. The most "advanced" cocci, equipped with flagella and spores (Sporosarcina), are aerobic and may therefore, have developed later than the photosynthetic lines.

The bacteria have been pictured as developing along five main lines (Kluyver and van Niel, 1936; Stanier and van Niel, 1941). These comprise (1) cocci; (2) the photosynthetic bacteria, which parallel lines (1) and (3); (3) polarly flagellated forms; (4) the largely non-motile gram-positive rods; (5) peritrichously flagellated rods, developing spores later. This scheme, like all such schemes, is highly tentative.

There are several difficulties. In the first line spores appear independently from those in the fifth line, in which spores and gram-positivity appear at the same time. In the second line, photosynthesis appears independently from that in the blue-green algae.

Here, too, it is perhaps unfortunate that the anaerobic photoautotrophs, probably appearing early in evolution, should be placed so closely parallel to the chemo-autotrophs (morphologically Pseudomonads), which are highly aerobic and likely to have occurred late in time. The bacterium and bacillus group (fifth line) is highly heterogeneous and also does not bring in the anaerobic forms first. However, it does begin with gram-negative and end with gram-positive forms. The fourth line is morphologically very satisfactory, but there are motile streptococci and Corynebacteria, the latter having *polar* flagella (Ferguson Wood, 1950). Even the Actinomycetaceae can form motile cells, which suggests that flagella arose more than once. However, a favourable feature is that both this and the third line become aerobic as they become advanced; in the Corynebacteria, *C. diphtheriae* and *C. acnes* are fermentative and the rest aerobic, so that this group is at the border line of aerobiosis.

Although Pseudomonas is aerobic, it forms unsaturated acids by the chemical mechanism used by anaerobes (Bloch *et al.*, 1962), which suggests close anaerobic relatives. The primitive coccus give rise to the Myxobacteria, but it is also possible (Stanier and van Niel, 1941) that the Myxobacteria arose from the blue-green algae by loss of chlorophyll, just as Beggiatoa and other sulphur bacteria are generally considered to have done.

The presence of creeping motility and absence of flagella in both groups suggest a relationship. Bisset (1955) points out that since peritrichously flagellate cells swim less actively than polar ones and can swarm over moist surfaces, they may constitute the first steps in the "adaptation of a fundamentally aquatic organism to a terrestrial environment." The polar types would be the more primitive, and it is noted that motile stages in the Caulobacteria and the Chlamydobacteria are also polarly flagellate. The polar and peritrichous groups are well separated, which suggests that the two kinds of flagella arose independently; also, the blue-greens branched off before flagella appeared. If anaerobic life and polar flagella were both primitive, it is perhaps significant that the *polar anaerobes* are not in evidence now. In other words, perhaps they were lost in evolution. The only polarly flagellated anaerobes existing now are in the photosynthetic group.

After adaptation to a terrestrial life, the next logical development would be that of aerial structures stalked fruiting bodies in the Myxobacteria, aerial hyphae and conidia in the Streptomycetaceae, and air-resistant spores in the Bacillaceae (Bisset, 1955). These adaptations to aerial life obviously arose independently. But independent multiple origin of organelles, such as spores, flagella, or photosynthetic structures, is difficult to avoid on any present scheme (Bisset, 1952).

Other workers have applied the criteria of (1) specific compounds, (2) nutritional requirements, (3) toxin formation, and (4) phage susceptibility to the evolutionary problem (Cowan *et al.*, 1955, and Symposium, 1962). As to (1), the appearance of hemoglobin in three unrelated groups, namely, vertebrates, yeasts, and legume nodules, does not give one much confidence in the idea. The second proves useful, however, in the Rhodobacteriinae and also in the Bacillaceae, where the general characters correlate well with nutritional patterns (Knight and Proom, 1950). The third has value in the Clostridia and the fourth in several groups. But these are taxonomic characters rather than guides to evolution.

Two new biochemical approaches to evolution and taxonomic relationships have recently become available. The first concerns the formation of lysine, which in all the bacteria studied (as well as in the green algae and higher plants) proceeds via the same intermediate, namely, diaminopimelic acid. The C-1 of alanine and the C-3 and C-4 of aspartate are readily incorporated into lysine, but C-1 of acetate not at all (Vogel, 1959). The Ascomycetes, Basidiomycetes, and the green Euglenidae form lysine by another route, in which C-1 of acetate is used. The Phycomycetes are divided between the two (Vogel, 1960-1961). This suggests that the fungi are not derived from the bacteria through the Actinomycetes, but rather form a separate line, arising independently from some of the Chytrids, a group which mycologists have long felt to be near the origin of the fungi.

The second approach, more general, rests on the base composition of the DNA's isolated from bacteria. First determined chemically by Chargaff and his associates (1950), and more recently by indirect, physical methods, the base composition shows striking relationship to the taxonomy.

Figures assembled by Belozersky and Spirin (1960) and Sueoka (1961) show the following percentages of guanine and cytosine:

Streptomyces (3 spp.) 74

Micromonospora 72

Pseudomonas (3 spp., 8 strains) 63-67

Azotobacter (3 spp., 6 strains) 55-57.5

Aeroh. aerogenes (6 strains) 55-59

Erwinia (7 spp.) 52-54

Salmonella (7 spp., 10 strains) 49-54

E. coli (10 strains) 50-53

Bacillus (3 spp., 11 strains) 34-46

Streptococcus (3 spp., 5 strains) 34-39

Clostridium (4 spp., 5 strains) 27-32

Each genus falls within a narrow range, and the interrelations (e.g., of the Enterobacteriaceae or the Actinomycetes) are obvious. *Aerobac. macerans* with G + C = 50 per cent is separated from other bacilli; the Sarcinae with 69 to 74 per cent seem more related to the Actinomycetes than to the Streptococci, and Neisseria with 50 per cent falls with the gram-negative rods, rather than with the gram-positive cocci. Within closer groups the relations hold good too, for *Bac. subtilis* with 43 per cent G + C will undergo genetic transformation only with bacilli having 42.5 to 44 per cent, not, e.g., with *Bac. laterosporus* having 40 per cent (Marmur *et al.*, 1963).

Isolated DNA's can also form hybrid helices together *in vitro* if very closely related, e.g., *Bac. subtills* and *Bac. natto,* both 43 per cent (Schildkraut *et al.*, 1961); *E. coli* strain B hybridises partially with Shigella and in one case with *E. freundii*, but not with Salmonella. This method may well help to give bacterial taxonomy a much firmer basis.

A point which may have evolutionary significance concerns the fixation of nitrogen. If combined nitrogen became as scarce as carbon compounds after the initial organisms had depleted the supply, then nitrogen fixation would have been an important character at the time when the synthesis of organic compounds was being developed. The ability to fix nitrogen is, suggestively, limited to the bacteria and, alone among the algae, to the lowliest group the blue-greens. In the bacteria, it is present in most of the anaerobes, both the clostridia and the photosynthetic forms, which we have given reasons for believing to be relatively primitive.

The possible relationship between Azotobacter and certain members of the blue-green algae also fits in well with this idea. Only the peculiar case of nitrogen fixation by symbionts remains out of place, but until we understand just what are the relative contributions of legume and Rhizobium, this is only to be expected.

It was suggested above that Pseudomonas, although aerobic, has close kinship with anaerobes, and the marked ability to fix nitrogen in this group supports that point. It has often been suggested that viruses represent primitive organisms. But it is clear that the property of carrying out metabolism is more primitive than that of self-reproduction. It is true that self-reproduction is the property most commonly associated with the idea of "organism," and that its development gave an almost infinite selective advantage to the organism that achieved it.

But the metabolism (probably fermentative) must have come first. Besides, the viruses are not "self-reproducing, but require the services of a more advanced form for reproduction. Therefore, it seems more likely that they represent a quite different, and later, line of development. Some workers believe that phylogenetic lines should be the basis for the working classification of the bacteria. It is evident from the above, however, that such a "natural" classification will come only slowly. In the Bergey system some presumably "natural" relationships are used, along with many artificial ones.

The system has been extensively criticised, especially by

Rahn (1937), Stanier and van Niel (1941), van Niel (1946), and Ferguson Wood (1950). Its excessive division into so-called genera and species, some separated only by differences that could be due to a single enzyme, led Wood to remark: "It is obvious that this work will have very limited value until all genera and species are pruned to a bare minimum." As far back as 1764 Linnaeus wrote: "A multitude of genera is a burden on the memory, to be lightened by system."

A practical difficulty in use of the manual is that organisms are constantly being reassigned, new genera created, and old ones dropped. While this is perhaps inevitable, as the field develops, it makes for a certain amount of confusion. Wald (1952) complained, in connection with similar practices in the taxonomy of vertebrates: The most important thing about a name, after all, is that it remain attached to the thing it designates.

One wishes that once a name had come into common use for an organism, it could be stabilised for the use of busy persons who want nothing but that each animal have a name. A broader criticism is directed at the whole policy of developing a taxonomy like that of higher plants. Van Niel (1946) argued that any attempt to set out groups of real taxonomic significance should be abandoned, and bacteriologists instead should "set themselves the task of developing a large number of keys, each one aimed at the segregation of particular groups on the basis of some salient character." Such groups could be quite artificial, such as "red-coloured bacteria," or "bacteria capable of decomposing cellulose," and would be purely for purposes of determination. This series of keys would enable all bacteria to be determined and would bring out interrelationships of many new sorts.

Sneath (1957) has proposed instead to compare pairs of organisms by supplying all possible recorded characters to a computer, which calculates a "similarity-value," S. The nearer S approaches 100, the more closely the two organisms are considered to be related.

The characters used have to be qualitative. Two organisms that have a recent common ancestor will necessarily have a high S- value, which gives the system some value for phylogeny (Floodgate, 1962) and may help in the problem of distinguishing important from unimportant characters.

Such fresh approaches may prove useful, but they do not afford much insight into the relationships. The difficult problems of bacterial classification are too complex to be discussed here, and it seems likely that a phylogenetic system will be a long time coming (Cowan *et al.*, 1955, and Symposium, 1962). The striking differences between the bacteria and the other orders, make one wonder whether bacteriology has as yet done any more than scratch the surface of the subject. That properties so totally different as chemosynthetic autotrophy, methane formation, and transduction should all be strictly limited to the bacteria is remarkable.

Perhaps for many physiologists this represents the fascination of the bacteria, that although they are so complexly interrelated, yet they comprise modes of life more varied than all the other plant types together. They offer an outstanding example of the evolutionary success that attends upon versatility. Perhaps it is because of this very versatility that, although they have been so intensively studied, they remain so profoundly unknown.

2

The Fermentations

Types of Fermentations

The action of microorganisms on carbohydrates falls naturally under four heads: The fermentation of simple sugars, The hydrolysis of diarid trisaccharides to simple sugars, The breakdown of complex polysaccharides and The synthesis of polysaccharides Since these processes typically do not require oxygen, they can all be regarded as fermentations. Of the fermentations of simple sugars there are six main types caused by bacteria and fungi. These can be classified as follows:

1. Alcoholic fermentation, in which sugar is converted to ethanol + CO_2.
2. Lactic fermentation, in which sugar is converted either to lactic acid alone, or to a mixture of lactic acid with alcohol or acetic acid.
3. Propionic fermentation, in which sugar (or lactic acid) is converted to propionic and acetic acids -f- CO_2.
4. Formic fermentation, a more complex type in which formic acid is the characteristic product, along with many other compounds in varying amounts.

5. Butyl-butyric fermentations, a still more complex group characterised by production of three four-carbon compounds, namely, butanol, butyric acid, and acetoacetic acid, usually together with alcohol and acetic acid.
6. The so-called "oxidative fermentations which are not fermentations at all but oxidation of the sugars and their breakdown products to various acids, and especially to acetic acid.

Of the six types the alcoholic fermentation is the most famous and also the most widely investigated. It is carried out par excellence by the veasts: also by some other fungi, when growing under anaerobic or semi-anaerobic conditions. A few bacteria give a modified alcoholic fermentation.

It will be simplest, therefore, to consider first the fermentation as it is carried out by yeast. The yeasts are a large group of fungi. They are normally single-celled, but some grow out into a limited mycelium. They include:

(1) The organisms forming ascospores and belonging to the family *Endomycctaceae;*

(2) Another group not forming ascospores (probably through having lost this ability), comprising the family *Torulopsidaceae;* and

(3) Several other genera which seem similar but may not actually be very closely related. The first group includes Saccharomyces, often referred to as "yeast"; the second group includes many which do not ferment at all but only oxidise.

Yeast cells may be diploid or haploid, but only diploid cells form asci, and even that is erratic and uncertain. Many media have been devised to promote ascus formation; high K, low N, addition of vegetable juices and acetate, all tend to promote the process. Light also promotes it in some yeasts, but inhibits in others (Oppenoorth, 1956). The controlling mechanisms remain unknown is near the optimum for alcoholic fermentation. The infection is spontaneous, and Pasteur was the first to show that yeasts are present on the surface of grapes especially as they approach maturity and form vigorous colonies at points of injury where the sweet juice exudes.

The yeasts are carried from one grape to another by insects, especially wasps. Wortmanri showed that carefully expressed sterile juice did not ferment until a wasp alighted on the solution. Among the yeasts on the surface of grapes and figs are:

(1) Saccharomyces, the principal alcohol-fermenting genus;
(2) Several of those types which fuse to produce ascospores, particularly Zygosaccharomyces species; and
(3) Nonsporeforming yeasts belonging to the genera Torulopsis and Candida.

The yeasts in the fermenting vats of wines are often *Saccharomyces "ellipsoideus"* a variety of *Sacch. cereviseae.* They convert the dextrose and fructose in the fruit juice into alcohol, up to a maximum of 14 per cent, at which concentration both fermentation and growth of the yeasts stops. Wines higher in alcohol than 14 per cent are made by distilling the fermented liquor and adding the distillate (grape brandy) to ordinary wine.

The flavour of wines is in large part due to the amyl alcohols and their esters, along with acetic and lactic acids and ethyl acetate. The acids are partly present in the fruit and partly formed in traces as by-products in the fermentation; the amyl and other organic radicals besides ethyl are produced in part by deamination of leucine and the other amino acids, in part synthesised directly.

Beer

Beer differs from wine in being made from seeds. Cereal seeds, including wheat, barley, rye, and corn, are allowed to begin germination, and then mashed and boiled with water. The resulting extract wort contains a moderate concentration of dextrose which has been set free from the starch in the grain by amylase. Since the amylase begins to act soon after soaking, and the resulting sugar is rapidly used up in growth and respiration of the seedling, careful control of the time and temperature of soaking is needed in order to control the sugar content of the wort.

Molasses or other forms of sugar may be added as well. Hops (the flowers of *Humulus lupulus* or, in America, *Humulus americand)* are also added, usually when the boiling is nearly over, both for their bitter flavour and because the extract has mild antiseptic properties, due to the resins humulon and lupulon. The antiseptics discourage the growth of Lactobacilli and Sarcinae, which occasionally damage the beer. The wort is cooled and at once inoculated, usually with *Saccharomyces cereviseae;* the fermentation produces not more than about 6 per cent alcohol.

The yeast is added in large quantities ("pitched") usually from a previous fermentation. Bottom yeasts, which grow as a sediment, and top yeasts, which grow mainly near the surface, carry out essentially the same fermentation, the former having a lower optimum temperature. Both are races of *Sacch. cereviseae.* The stability of the froth ("head retention"), colour, and flavour are largely controlled by the nitrogen compounds in the wort. In dark ales and stout, a slow fermentation carried out by *Brettanomyces* species follows the main one, giving additional flavour and darker colour. Molasses and roasted grain may be added for darkening.

Industrial Alcohol

This product is usually made from molasses (as a cheap source of sugar), in which the sucrose is rapidly hydrolysed by invertase in the yeast. In recent years acid-hydrolysed sawdust has been used on an experimental scale, and sulphite waste liquor from paper factories is also sometimes fermented (Kolachov and Nicholson, 1951). As far as possible, pure cultures of *Sacch. cereviseae* are used. In wine and beer fermentations, "wild yeasts," of the family *Torulopsidaceae,* cause trouble by producing bitter tastes, but in industrial alcohol production their effect is to reduce the yield of alcohol, because these are mainly aerobic forms which oxidise away the carbohydrate. To obtain as nearly pure culture Saccharomyces as possible, small pure cultures are used to inoculate successively larger vessels (as done originally by Hansen), the last being the so-called "propagation plant" in the distillery. It may be as large as 7000 litres, so that 5 per cent of the fermentation tank itself (30,000 gal) is supplied by this inoculum. In using molasses, urea is often added, since the nitrogen content of the molasses is too low to allow really vigorous yeast growth. An acid *pH.* (about 4.7) is maintained to suppress bacteria.

Bread

Essentially this is also an alcoholic fermentation product. The source of sugar is the small amount in flour, plus a little which is usually incorporated in the dough. The fermentation is of short duration (4 to 12 hours) and is intended to produce gas to "leaven" or *lift* the dough. The French word for yeast *(levure)* and the old English "leaven" refer to this use of yeast. The alcohol is, of course, driven off in the baking. Baker's yeast is a "top yeast" form of *Sacch. cereviseae.*

In ancient days bread and beer were made together from wheat flour. The models of the life and times of 2000, BC, excavated at Thebes in Egypt, show *(inter alia)* a brewery and bakery in one building. "In the first room two women grind corn into flour and a man makes it into cakes of dough, which another treads into mash in a barrel. Nearby, the rising mash stands in four tall crocks while the yeast ferments, another man pours it into a row of stoppered jugs which stand along the wall" (Winlock, 1921). In the next room is the bakery, doubtless being supplied with yeast from this mash. This is probably the oldest known representation of a microbial process.

Yeast

Yeast itself is produced in quantity when the medium is aerated. The yield of alcohol decreases, and the yeast grows much more than under anaerobic conditions. A cereal mash, as for beer, is used, glucose being added to increase the yield. In recent years ammonium salts have been used as a main source of nitrogen; the yeast converts this in 50 per cent yield to protein.

Hence this "grainammonia process" is more efficient than agricultural methods, like pasturing cattle, for the formation of concentrated protein, while the resulting yeast cake is at least as good a food as beef. When sugar, ammonium, and phosphate are added steadily as the yeast grows, yields as high as 90 per cent, calculated on the nitrogen, are obtained. The effectiveness of grain as basal medium is due to the need of yeast for vitamins as growth substances. These compounds, formerly known collectively as "Bios," and later separated into Bios I, II, III, and so on, are present in cereal seeds. They comprise:

(1) Myo-inositol, $CH_{14}O$ or hexahydroxycyclohexane.
(2) Biotin, a compound containing nitrogen and sulphur.
(3) Nicotinic acid, or its amide.
(4) Thiamine.
(5) Riboflavin $CH_2(CHOH)_3CH_2OH$ or vitamin.
(6) Pantothenic acid, which comprises part of the molecule of co-enzyme A, and is a peptide of pantoic acid and 3-alanine. The moleculus formula of Pantothenic acid in $C_9 H_{17} NO_5$. Some yeasts do not require the whole molecule but can synthesise it if supplied with alanine, or even with peptides of alanine (Betz, 1960-62).
(7) Folic acid, which contains the pteridine nucleus, linked with p-aminobenzoic acid and up to 7 molecules of glutamic acid.

The functions of these substances are more or less understood. Nicotinic acid and riboflavin are converted to co-enzymes and flavoproteins. Biotin catalyses carboxylation reactions; thiamine, converted to its pyrophosphate, is the co-enzyme for carboxylase; pantothenic acid is converted to co-enzyme A; and folic acid is required for certain one-carbon syntheses.

All yeasts do not require all the vitamins, and a few can grow on synthetic medium; one of these, *Torulopsis utilis,* was proposed as a food during World War I, being raised on hydrolysed sawdust and ammonium salts. It can also be raised on sulphite waste liquor from paper mills after this has first been fermented by Saccharomyces.

The Saccharomyces attacks only the hexoses, and Torulopsis then grows on the pentoses; urea and phosphate are added to the liquor as nutrients. In this way a ton of waste will yield 20 gal of alcohol and 40 to 100 lb of yeasts, which are used as high-protein cattle feed (Kurth, 1946).

Early Work on the Alcoholic Fermentation

The nature of fermentation was always part of the "spontaneous generation" controversy. It was Schwann, in 1837, who showed that the alcoholic fermentation of grape juice is entirely dependent on germs reaching it from the atmosphere.

From the morphology of yeast, and its susceptibility to different poisons (Hg" and arsenate, but not to strychnine, which poisons mammals), Schwann deduced that yeast belongs to the plants rather than the animals. Cagniard-Latour and Kiitzing in the same year also made quite clear that yeast is a plant.

However, Liebig and other chemists had always held that fermentation is due to the presence of unstable molecules, which in their breaking down carry along other molecules, thus causing them to break down also. The need for a trace of yeast was thus viewed as merely the need for what we should now call a catalyst, i.e., (in Liebig's view), some unstable matter in the process of breakdown. In 1828 Wohler synthesised urea, a typical biological product, and thus he and Liebig were satisfied that in the end all biological processes would prove to be chemical in nature. In alcoholic fermentation, they were supported in their contention by the simple molecular relationships expressed in the Gay-Lussac equation:

$$C_6H_{12}O_6 \rightarrow 2\ C_2H_5OH + 2\ CO_2$$

This equation satisfies the law of simple proportions and thus gives a deceptively "chemical" appearance to the fermentation. Actually, as Pasteur showed, it does not truly describe the fermentation, because small quantities (fractions of a mole) of glycerol and succinic acid also appear. The biological concept of fermentation as something resulting from the life of microorganisms, therefore, seemed absurd to Liebig. He and Wohler published an anonymous paper, mocking at the idea that living yeast could be responsible for fermentation, and at the use of the microscope in what they regarded as purely chemical studies. This appeared in Liebig's journal, the *Annalen*, in 1839. A free translation of part of it is as follows:

> I am in process of developing a new theory of wine fermentation . . . This discovery shows how simple are the means used by Nature to cause the most wonderful phenomena. I owe this to the application of the microscope .... Beer yeast broken up in water is resolved by this instrument into innumerable small spheres. When placed in sugar water it can be seen that these are the eggs of animals; they swell, burst, and there develop small animals which multiply with incredible rapidity in a most unprecedented way.

The form of these animals differs from that of the 600 spp.; it is the shape of a Beindorfs distilling flask (without the condenser). The tube of the stillhead is a kind of sucking snout covered internally with fine 1/2,000-line cracks; although teeth and eyes are not to be seen, one can distinguish a stomach, intestine, the anus (a rose-pink spot), and the organs of urine secretion. From the moment of emergence from the egg, the animals suck in sugar, which can clearly be seen in the stomach. It is immediately digested and the digestion is followed by excretion. In a word these infusoria feed on sugar; they excrete from the intestine alcohol and from the urine organs carbon dioxide. The urine bladder in the full condition is shaped like a champagne bottle. This paper may have had considerable influence in delaying the further study of fermentation and of spontaneous generation. At any rate it was 20 years before Pasteur elaborated and repeated Schwann's experiments. He confirmed that there is no spontâneous generation, and that fermentation is due to the living yeast (Pasteur, 1860):

> I am of the opinion that alcoholic fermentation never occurs without simultaneous organisation, development, multiplication of cells, or the continued life of cells already formed. The results expressed in this memoir seem to me to be completely opposed to the opinions of Liebig and Berzelius.
>
> If I am asked in what consists the chemical act whereby the sugar is decomposed and what is its real cause, I reply that I am completely ignorant of it. He thus took the "vitalist" position that fermentation is inseparable from life. He considered that the organism, deprived of oxygen from the air, draws its oxygen from organic compounds: "*La fermentation est la vie sans l'air.*"

Shortly before Pasteur, Moritz Traube, in 1858, put forward the theory that *all fermentations produced by living organisms are in reality caused by ferments in the organism.* Ferments were considered definite substances *closely related to proteins.* This incredibly accurate prophetic view was supported when Berthelot, in 1860, isolated invertase from yeast, but there was no experimental support forthcoming for actual *fermentation* by cell-free products.

Pasteur himself and many other workers, including Claude Bernard, the physiologist, tried grinding, freezing, and plasmolysing yeast, and even extracting it with glycerol, without success. It was in 1897 that Buchner, grinding yeast with sand to make an extract for therapeutic purposes, found that the extract decomposed rapidly and attempted to preserve it by adding sugar.

The mixture gradually frothed, CO_2 was evolved, and cell-free fermentation was discovered. The papers of Buchner (1897) and Buchner and Rapp (1897-1899) created a sensation. Liebig had been half right after all. Later, numerous fermenting preparations were made from yeast. *Zymin* is made by treating yeast with alcohol and ether, or with acetone, and drying.

Lebedev made "maceration juice" by drying the yeast and then macerating it in water at 35° C. All such preparations ferment, although only from 2 to 10 per cent as fast as living yeast; they also give a slow "autofermentation" due to the breakdown of stored glycogen which is hydrolysed to sugar by the killed preparation. They ferment glucose, fructose, and mannose, also sucrose and maltose; raffinose is attacked very slowly. Preparations from yeasts other than *Sacch. cereviseae* may attack galactose. Pentoses are not fermented by these preparations, but phosphorylated pentoses are, and Lebedev juice ferments ribose phosphate as fast as glucose (Dickens, 1938).

Role of Phosphate

So long as the living cell is required, research on biological processes is slow. In fermentation, as soon as cell-free preparations could be used, a major discovery was made which opened up a whole new vista in biological chemistry. Buchner in 1903 observed that CO_2 production was stimulated by adding phosphate and attributed this to the increased acidity of the medium. It was in a more careful study of the influence of phosphate that the major discovery referred to was made, by Harden and Young (1906).

They found that the addition of phosphate to fermenting

yeast juice caused the rate to rise by as much as 20 times for a while; it then returned to normal. On adding a second quantity of phosphate the rise was again obtained. Each addition caused production of one additional mole of CO_2 per mole of phosphate added, and alcohol was also produced in proportion to the CO_2. A smaller effect was obtained with zymin, but none with intact yeast, probably because the cells contained enough phosphate already.

The full explanation of this effect of phosphate took about 40 years to elucidate. When at the end of the experiment the liquid was examined, the phosphate was no longer free as HPO_4 ions, but as organically bound phosphate *esters*. They isolated a hexosediphosphate, afterwards found to be fructose (fructofuranose)-l-6-diphosphate. By careful hydrolysis of this, Neuberg (1918) subsequently prepared a fructose (fructofuranose)-monophosphate, while Robison (1922) isolated from fermentation liquor a third ester, glucose (glucopyranose)-6-phosphate. Subsequently another compound, glucose-1-phosphate, intermediate in the breakdown and resynthesis of starch and glycogen, was identified by Cori and Cori:

> Altogether, no less than 16 phosphate compounds participate in fermentation, five as hexose esters, eight as intermediate substrates, and three as co-enzymes. The old slogan, *"Ohne Phosphor kein Leben"* thus receives its full explanation.

Co-enzymes

When yeast juice is thoroughly dialysed, then neither the dialyzate (small molecules) nor the residue (large molecules) can carry out fermentation, but on mixing them together they ferment as before. This discovery was made independently by Buchner and Antorii and by Harden and Young, both in 1905. The residue, called *apoenzyme,* is inactivated by heat and thus proteinaceous, while the dialyzate, called *co-enzyme* or *cozymase,* is thermostable.

The cozymase cannot be replaced by phosphate ions, but is organic in nature. In this connection a discovery of great importance for comparative biochemistry was made by Meyerhof (1918). He found that when animal muscle is ground up and thoroughly washed, it becomes unable to carry out its typical fermentation, namely, the formation of lactic acid from carbohydrate. If, however, the washings are added to the residue, the mixture is again able to form lactic acid. This substance is closely related to co-enzyme II; it differs only in containing but two phosphoric groups while in the other, called triphosphopyridine nucleotide, TPN, or NADP, the C atom marked with an asterisk is also phosphorylated. However, their actions are quite specific, each working most efficiently with its own group of dehydrogenases.

Thiamine Pyrophosphate, Diphosphothiamine, or Cocarboxylase, the CO_2-releasing Co-enzyme (Lohmann and Schuster, 1937): Yeast cocarboxylase, if carefully purified at *pH* 6, contains the diphosphothiamine and also magnesium attached to the protein, in a ratio of 1 Mg atom and 1 diphosphothiamine molecule for 75,000 gm of protein (Kubowitz and Liittgens, 1941). At *pH,* 8 the co-enzyme dissociates off, leaving the apo-carboxylase, but the whole enzyme, *carboxylase,* can be resynthesised in presence of a metal ion and pyrophosphate, and its activity fully restored (Boffi *et al.,* 1959). Yeast incubated with thiamine and phosphate synthesises first the cocarboxylase and then thiamine triphosphate, which is inactive (Kiessling, 1956-1959). Cell-free extracts of *My cob. lacticola* do the same thing (Rossi-Fanelli *et al.* 1960); however, they can also split off the phosphate groups again, so that the net synthesis soon stops. In addition to the organic co-enzymes, the ions Mg^{++}, Mn^{++} and NH_4^+ are also essential for the fermentation.

Reducing Power of Yeast

The next important step in the understanding of the fermentation is the appreciation that yeast, particularly in the presence of sugar, constitutes a strongly reducing system. Fermenting yeast can reduce powdered sulphur, methylene blue, or thiosulphate. Yeast which is fermenting sugar can also reduce aromatic nitro compounds to amines, ketones to secondary alcohols, and of particular importance aldehydes to primary alcohols. Most notable is the great readiness with which acetaldehyde is reduced to ethyl alcohol (Kostytschev and Hubbenet, 1912-1913), a reduction carried out equally well by living yeast, pressed dried yeast-cake, or zymin. The significance of this is seen in connection with the fundamental theory that acetaldehyde is an intermediate product in the fermentation.

Acetaldehyde, an Intermediate Product

It was Neuberg and co-workers who in 1911-1912 discovered that yeast decomposes keto acids to CO_2 and aldehydes.

Oxalacetic and α -ketoglutaric acids are also decomposed, as well as a number of the keto acids formed by oxidative deamination of amino acids (Meister, 1952). Especially vigorous is the decomposition of pyruvic acid:

$$CH_3COCOOH \longrightarrow CH_3CHO + CO_2$$

The rate of production of CO; from pyruvic acid is faster than the production of CO_2 from glucose itself. The enzyme concerned, *carboxyjase,* is widely distributed in yeasts, fungi, and higher plants It has a *pH.* optimum between *pH* 4 and 5, and requires the diffusible co-enzyme, *cocarbozylase,* or diphosphothiamine, described above. The presence of carboxylase, and the ease with which acetaldehyde is reduced by yeast, led both Neuberg and Kostytschev to conclude that fermentation proceeds via the formation and decarboxylation of pyruvic acid and subsequent reduction:

$$\text{Hexose} \longrightarrow CH_3COCOOH \longrightarrow CH_3CHO$$
$$\longrightarrow CH_3CH_2OH + CO_2$$

Thus, the CO_2 of fermentation comes from the carboxyl

group of pyruvic acid. This was the first instance of the general rule that all CO_2 produced biologically comes from carboxyl groups. If $CaCO_3$ be added during fermentation of glucose, calcium pyruvate is precipitated. The best evidence, however, is given by Neuberg's fixation process, or *Abfangverfahren.* This consists in adding sodium bisulphite to a fermenting mixture of yeast and sugar. The acetaldehyde-sulphite compound is precipitated, and *glycerol,* accumulates.

The production of glycerol from sugar attracted great interest, because no sooner had it been worked out than Germany found herself in World War I, with a drastic shortage of glycerol, due to the lack of imported fats. By 1917 it is estimated that the process was making over 1000 tons of glycerol per month, giving a 15 to 20 per cent yield on the basis of sugar fermented. Some writers have claimed that without this process the Germans would not have been able to wage war after the end of 1917. In place of sulphite, two other substances that combine with acetaldehyde, namely, dimedon and semicarbazone, have been used in later experiments with similar results. With *Sacch. rouxii* oxygen has the same effect (Spencer and Shu, 1957).

This yeast forms arabitol as well as glycerol. The fermentation has also been carried out with yeasts "adapted" to the presence of magnesium sulphite (Underkofler *et al.*, 1951). Not only does the formation of acetaldehyde-bisulphite prove the theory, but the appearance of the glycerol proves some point else. One mole of glycerol appears for every mole of acetaldehydebisulphite formed. Had the acetaldehyde not been removed, it would have been hydrogenated to alcohol. Hence the glycerol must contain the hydrogen, which would have been accepted by acetaldehyde. The only compound that could be hydrogenated to glycerol. This triose actually consists of a mixture of the phosphorylated derivatives of glyceraldehyde and dihydroxyacetone: These two substances are in equilibrium with each other through the enzyme isomerase. (Under physiological conditions the equilibrium ratio is about 95 per cent phosphodihydroxyacetone.) Neuberg considered the normal fermentation the *first form*, and the bisulphite-glycerol fermentation the *second form*. He also discovered a *third form of* fermentation, occurring in the presence of weak alkali. Here acetic acid is produced, and the yield of alcohol falls to half the normal. This process is *dismutation;* two molecules of acetaldehyde react together to form one of alcohol and one of acetic acid:

$$CH_3CHO + CH_3CHO \longrightarrow CH_3CH_2OH + CH_3COOH$$

The reaction is catalysed by two enzymes, aldehyde dehydrogenase and alcohol dehydrogenase, linked by DPN (Racker, 1949). Since acetaldehyde is thus removed and cannot act as hydrogen acceptor, 4th cerol accumulates here too. For each mole of acetic acid, two moles of glycerol appear. This reaction has been considered for industrial glycerol manufacture, but the fermentation slows down markedly in alkali. Increasing the glucose concentration to 25 per cent also increases the formation of glycerol, giving yields above 30 per cent of the glucose fermented.

Early Stages of Hexose Breakdown

Understanding of the first stages, which were for a long time puzzling, only became possible by the recognition that both hexoses and trioses react in the form of their *phosphate esters.* The recognition of the phosphotrioses is due to Meyerhof and Kiessling (1933-1935) and to Barrenscheen (1933), who showed that phosphoglyceraldehyde is fermented at least as fast as glucose.

Synthesis of (racemic) phosphoglyceraldehyde is due to Fischer and Baer (1932), and its isolation to Meyerhof and co-workers. The two phosphotriose molecules, phosphoglyceraldehyde anophosphodihydroxy acetone, are formed from one diphosphofructose molecule by the enzyme *aldolase.* This aldolase reaction is reversible (Warburg and Christian, 1943) which is very important for synthesis, since it means that hexose can be formed directly from triose. The phosphoglyceraldehyde first takes up a molecule of inorganic phosphate and at the same time is dehydrogenated to 1, 3-diphosphoglyceric acid, or phosphoglyceryl phosphate.

The hydrogen is accepted by DPN (cozymase, NAD), the hydrogen-transferring enzyme being *phosphoglyceraldehyde dehydrogenase.* This complex reaction was demonstrated very elegantly by Warburg and Christian (1939), who called it the "oxidising reaction of fermentation." If phosphoglyceraldehyde is mixed with DPN and crystalline phosphoglyceraldehyde dehydrogenase (which they isolated), the DPN is not reduced unless inorganic phosphate is added.

The rate of formation of reduced DPN varies with the phosphate concentration; arsenate will shift the equilibrium drastically. These phenomena are explained by the nature of the aldehyde dehydrogenase enzyme, which contains both two reactive SH groups and two molecules of the co-enzyme DPN (Racker and Krimsky, 1952; Velick *et al.*, 1953). The phosphate linkage to a double-bonded carbon atom in 1,3-diphosphoglyceric acid is "energy-rich"; it would not have been formed in appreciable concentration by phosphorylation of 3-phosphoglyceric acid, since its formation requires about 10,000 calories.

This energy is furnished by the dehydrogenation of the phosphoaldehyde (Meyerhof and Oesper, 1947). Because of the high bond energy, diphosphoglyceric acid can phosphorylate a phosphate acceptor, which is adenosine diphosphate, converting this to the triphosphate, ATP. The enzyme responsible is called *phosphoglycerie phosphokinase.* The resulting dephosphorylated compound is 3-D-phosphoglyceric acid.

Hence the overall change is that the inorganic phosphate which was taken up in the first step is converted to ATP, while the phosphoglyceraldehyde is dehydrogenated to the acid. This compound was first identified in an alcoholic fermentation proceeding in presence of fluoride (Nilsson, 1930). The overall reaction: 3-Phosphoglyceraldehyde + DPN[4] 3-Phosphoglyceric acid + DPNH -f H[4] yields 7500 to 10,000 calories (Meyerhof and Oesper, 1947) in going from left to right. The 3-phosphoglyceric acid now undergoes a curious set of changes (Sutherland, Posternak, and Cori, 1949). A small amount of it is rephosphorylated, probably by ATP, to form the 2, 3-diphosphoderivative. (Both phosphate groups in this are ordinary esters linked by low-energy bonds.) The diphosphocompound then gives up the phosphate group in the 3 position to phosphorylate the 2 position of another molecule.

Thus, a new molecule of 2-3-diphosphoglyceric acid is formed, which can now react with a second molecule of the 3-phospho acid. The net result is conversion of one molecule of 3-phospho acid to a molecule of the 2-phospho acid. The 2,3-diphospho acid is acting as a co-enzyme. The enzyme catalysing this reaction *(phosphoglyceromutase)* is thus a phosphate-transferring enzyme or "trans-phosphatase." The resulting 2-phosphoglyceric acid loses a molecule of water under the influence of another enzyme, *enolase,* to produce phosphoenolpyruvic acid, i.e., pyruvic acid in the enol form, phosphorylated at the OH group.

This reaction is comparable to the conversion of malate to fumarate by fumarase. It is of special interest because of its sensitivity to poisoning by fluoride.

Enolase requires magnesium ions for its activity, and in presence of fluoride a magnesium fluorophosphate is formed which combines with the protein enzyme and inactivates it (Warburg and Christian, 1942).

As a result, 3-phospho and 2-phosphoglyceric acids accumulate and could be isolated by Nilsson and by Meyerhof, respectively. The next slep is the dephosphorylation of the phosphoenolpyruvic acid. Like the 1 -phosphate bond in 1,3-diphosphoglyceric acid, the phosphate bond here is linked to a doubly bonded carbon atom and is therefore energy -rich. As before, it phosphorylates ADP, which as a result is converted to the triphosphate.

Thus, in these steps two molecules of ADP have been converted to ATP. Since both energy-rich phosphate bonds retain their energy in ATP, they can react with other substrates and phosphorylate them. In this system they react with glucose, in presence of the enzyme *hexokinasc,* to produce first the 6-phosphate, then fructose 1,6-diphosphate (Harden-Young ester).

Thus, the cycle is started again. A little of the phosphoenolpyruvate suffers a different fate; it combines with CO_2. The phosphate is split off at the same time, and oxaloacetate results. This is then reduced to malic and succinic acids or can be transaminated to aspartic acid.

These reactions, which account for a group of minor products, are inhibited by oxygen (Stoppani *et al.,* 1958), and hence the amounts of these acids vary with the conditions. In the formation of phosphoglyceric acid, a molecule of DPN has been reduced. The reduced DPN has two alternative fates. At first it is dehydrogenated by phosphodihydroxyacetone to produce 3-phosphogly cerol.

Which is hydrolysed to glycerol by a phosphatase. It is in this way that small amounts of glycerol are formed in the normal fermentation (Pasteur).

Later, however, acetaldehyde appears, and this is much more reducible than phosphodihydroxyacetone. Therefore, from now on, whenever phosphoglyceraldehyde is dehydrogenated, the 2(H) resulting is taken up eventually by acetaldehyde. The role of DPN in carrying 2(H) from triose to acetaldehyde is parallel to that of ATP in carrying phosphate from triose to hexose.

Some Additional Points

It remains to explain Harden and Young's initial finding that the fermentation by yeast juice, with added phosphate, comes to a stop, with accumulation of hexosediphosphate. This is due to the peculiar mechanism of the reaction from phosphoglyceraldehyde to phosphoglyceric acid. Here the intermediate, 1,3-diphosphoglyceric acid, is produced by uptake of inorganic phosphate.

Since the 1-phospho group has high energy, however, it phosphorylates ADP. Inorganic phosphate is thus continuously removed from the system and transferred to ATP which at once phosphorylates hexose with it. Hence hexosediphosphate accumulates, and the system comes to a standstill for want of ionic phosphate.

If arsenate is added, phosphorylation is interfered with because, instead of phosphoglyceryl phosphate, phosphoglyceryl arsenate is formed, and this hydrolyses spontaneously to phosphoglyceric acid (Warburg and Christian, 1939).

The same result could be secured (Harden and MacFarlane, 1930) by adding a phosphatase preparation. In the living cell ATP is continuously being hydrolysed by a pyrophosphatase (ATP-ase), but this enzyme is too unstable to survive the grinding of yeast into juice. Indeed, it is destroyed by autolysis, slow drying, or acetone treatment; i.e., by the very treatments used for most biochemical studies (Meyerhof, 1949).

If yeast is quickly dried, and then broken by vibrating supersonically, its ATP-ase is largely retained. After washing, it needs the addition of DPN, ATP, phosphate, Mg^{++}, and either K^+ or NH_4^+, and will then ferment glucose at a steady rate (Meyerhof and Kaplan, 1951). Hexosephosphates definitely do not accumulate; phosphate is neither absorbed nor liberated to an appreciable extent.

The CO_2 contained about 3 per cent of the activity, probably owing to a small exchange of phosphate between the 1 and 3 positions of dihydroxyacetone. Yeast produces a small amount of alcohol and CO_2 when simply suspended in water.

This "autofermentation" is at the expense of glycogen stored in the cell. The splitting of glycogen is a phosphorolysis, brought about by ATP, and yields glucose-1-phosphate. This compound enters the fermentation by being converted to glucose-6-

phosphate, which involves a system exactly comparable to that converting 3-phosphoglyceric to 2-phosphoglyceric acid.

The enzyme is called *phosphoglucomutase,* and its co-enzyme is glucose-1,6-diphosphate (Leloir *et al.* 1948-1949). Small amounts of this diphosphate are formed from glucose-1-phosphate, using ATP, by the enzyme glucose- 1-phosphate-kinase. The diphosphate can then convert the 1-phosphate to the 6-phosphate.

Phosphoglucomutase requires Mg and, like enolase, is readily poisoned by fluoride. Its reverse action leads to glycogen synthesis, but another enzyme, using uridine-diphosphoglucose, is more important for this synthesis and probably controls it (Madsen, 1963).

Alcoholic Fermentations by Bacteria

Some bacteria produce alcohol from sugar. In several cases, this involves routes differing from that followed by yeast. The best example is *Ps. lindneri* (formerly *Zrmononas mobilis)* which, unlike other Pseudomonads, is an obligate sugar fer-nenter. Its fermentation produces the Mexican alcoholic drink, "pulque," made from the juice of Agave plants. *Ps. lindneri* ferments glucose or fructose (not mannose) to 45 per cent CO_2, 45 per 'cent ethanol, and up to 7 per cent of lactic acid (Kluyver and Hoppen- brouwers, 1931).

When glucose labelled with C^{14} in the 1 position is fermented by this organism, practically all the activity is found in the CO_2 (Gibbs and de Moss, 1951). Now in the normal breakdown of hexosediphosphate, it is carbon atoms 3 and 4 that produce the aldehyde groups. Thus, the alcohol represents carbon atoms 1 and 2, 5 and 6. With *Ps. lindneri,* glucose with C^{14} in the 3 and 4 positions yields half its C^{14} as CO_2 and half as ethanol, i.e., CO_2 comes from positions 1 and 4 and hence ethanol from 2 and 3, 5 and 6 (Gibbs and de Moss, 1954).

"Thus," says Kluyver (1956), "the Mexican who enjoys his alcohol as produced by *Ps. lindneri* imbibes different carbon atoms of the sugar than does the American who takes it in the form of Bourbon whisky." The key to understanding this fermentation was furnished by an aerobic organism, *Ps. saccharophila,* which produces pyruvic acid and triose. In this case the first step is conversion of the glucose-6-phosphate, not to a diphosphate, but to 6-phosphoglucomc acid.

Final formation of alcohol liberates carbon atoms 1 and 4 as CO_2. This explains the distribution of carbon atoms found for *Ps. lindneri* by Gibbs and de Moss; evidently, therefore, this organism uses the same pathway but carries the process all the way to alcohol and CO_2. A peculiarity of this fermentation is that one of the two molecules of pyruvic acid arises directly from the hexonic acid and not via phosphoglyceraldehyde. As a result no *ATP* is produced in its forma*tion.* The second molecule arises via triosephosphate, and therefore ATP is produced as with yeast. Thus, the whole process yields only one ATP instead of two.

The alcohol produced by heterofermentative lactic bacteria such as Leuconostoc arises in a somewhat different way; these bacteria produce lactic acid and alcohol. Here again glucose-1-C^{14} gives all its C^{14} in the form of CO_2, while glucose-3, 4-C^{14} gives half its C^{14} in the lactic acid and half in the ethanol (de Moss *et al.,* 1951).

Phosphogluconic acid is the first intermediate, as above, but apparently this is decarboxylated to a phosphopentose, which is split, carbon atoms 1 and 2 becoming ethanol, while numbers 3, 4, and 5 yield lactate.

Correspondingly the dehydrogenated product acetate appears. As the hydrogen content decreases, through gluconate to ketogluconate, the ethanol continues to decrease and the acetate to take its place. The exact reactions leading to alcohol are obscure. Another peculiarity is that the intermediate can hardly be an ordinary phosphopentose, since neither Leuconostoc nor *Ps. lindneri* can attack any of the pentose phosphates (deMoss, 1953).

The alcohol formed in the complex fermentations of the

coliform bacteria arises by yet another route, namely, from reduction of acetyl-co-enzyme A.

The giant sarcina, *Zymosarcina ventriculi,* making alcohol, acetic acid, CO_2, and Hy, may operate similarly. Alcohol is also formed, along with traces of butanol and butyric acid, by a strain of *Cl. tetani* (Lerner and Pickett, 1945), but the pathway has not been studied. The bacterial alcohol fermentations are thus based generally on hexose monophosphate instead of the diphosphate, and yield half as much ATP as does the yeast fermentation.

Inhibition of Fermentation by Oxygen: Pasteur Effect

Pasteur discovered that in air the growth of yeast was accelerated, while the consumption of sugar was reduced. In air the yeast (in his own words) "ceases to be a ferment." With intermediate amounts of oxygen the rate of sugar consumption was decreased proportionately. Evidently two phenomena are to be distinguished.

The second phenomenon is more obscure; clearly oxidation in some way slows down the attack on sugar. The phenomenon is shown not only by yeasts and their relatives but also to varying extents by Lactobacilli and Propionibacteria; it is found also in animal tissue, especially retina, and in higher plants. At least five explanations have keen advanced but none of them has been proved. One of the most widely accepted is that oxidation leads to such efficient phosphorylation that all the adenylic acid and ADP in the, aerobic cell are converted to ATP, and thus not enough ADP remains to accent the phosphate from diphosphoglyceric and phosphoenolpyruvic acids.

Sugar utilisation is thus decreased because of a shortage of ADP. In support of this idea, Meyerhof and Fiala (1951) found that, with the quickly dried yeast, nitrophenol decreases the production of CO_2 anaerobically and increases it aerobically, i.e., it prevents the Pasteur effect. Since nitrophenols are known to "uncouple" phosphorylation from oxidation, i.e., to prevent phosphate from being converted to ATP, it can be concluded that the Pasteur effect is prevented by decreasing the ATP formation.

An attractive and simpler view depends on the ready oxidisability of phosphoglyceraldehyde dehydrogenase. Admittedly the critical work has been done on this enzyme in pea-seed extract rather than in yeast, but the pea-seed extract shows a typical Pasteur effect (Hatch and Turner, 1959, 1960). Phosphoglyceraldehyde dehydrogenase depends for its activity.

In air this group oxidises to the S-S form, which is inactive. Hatch and Turner found that within an hour in air the total content of protein in the extract fell by a third, and phosphoglyceraldehyde dehydrogenase activity by 70 per cent. The oxidation is catalysed by a metal. Addition of DPNH or of a mixture of hexose phosphate with DPN reactivated the enzyme; in the second case the hexose phosphate was being metabolised to reduce the DPN. It is clear that if phosphoglyceraldehyde is not dehydrogenated to phosphoglyceric acid, it will accumulate in the form of hexose phosphate (since the aldolase reaction is readily reversible), and thus oxygen, by oxidising prevents the breakdown of sugar. In view of the lessened yield of ATP under anaerobic conditions, (one or two per hexose instead of 32 in air) the increased rate of sugar consumption represents a most appropriate control mechanism.

Lactic Bacteria

The Group as a Whole: The thorough studies of Orla-Jensen (1919) made clear that the true lactic acid bacteria are gram-positive, non-motile, and non-spore- forming, and include both spheres and rods. Except for some strains of *L. plantarum* they do not usually contain catalase, and consequently many of them produce H_2O_2 when growing in air. Three groups produce lactic acid from sugar in 80 to 98 per cent yield (in acid media), with only traces of by-products. These *homofermentative* forms include: *Thermobacterium:* long rods, optimum temperature about 40° C or higher, making either inactive or laevolactic acid in concentrations up to 3 per cent (0.33 M).

Streptobacterium: Chain-forming rods, optimum temperature about 30° C, making inactive or dextrolactic acid up to about 1.5 per cent. *Streptococcus* chain-forming spheres with various optimum temperatures, making dextrolactic acid in concentrations below 1 per cent. Some are haemolytic. *Sc. pyogenes* is not strictly a "lactic" organism at all, in the special sense that it will not grow in milk.

A few strains of cocci, called Tetracoccus or *Pediococcus* (Pederson *et al.*, 1954) form tetrads instead of chains; they are often called Sarcinae but are evidently related to the lactic Streptococci. Three other groups are *heterofermentative,* i.e., produce a mixed fermentation in which only about half the sugar is converted to lactic acid, the remainder appearing as CO_2, H_2, alcohol, formic or acetic acid.

Betabacterium: Growing on vegetables or cheese, but not milk, making inactive lactic acid (this genus name is no longer used). *Betacoccus* (or *Leuconostoc):* growing on ensilage and other plant material, in which they attack pectins and the resulting pentoses; when the ensilage is fed to cattle, the organisms appear in the milk. These bacteria are noted for growth in sucrose solution with formation of much slimy polysaccharide. The "frog's egg bacterium" *(Leuconostoc mesenteroides)* often makes trouble in sugar refineries for this reason. *Bacterium bifidum:* a strict anaerobe showing branched and forked forms, found in the intestinal flora of milk-fed infants. This and a related similar organism are probably anaerobic Corynebacteria, being of very variable morphology (Hayward *et al.*, 1955). They produce lactic and acetic acids, without CO_2 Several other anaerobic branched forms occur in faeces or tissue fluids (Prevot, 1938). The major properties of the lactic organisms, which is condensed from Orla-Jensen, Henneberg, Bergey, and others. In addition to the true lactic acid bacteria, milk, cream, and cheese often contain members of the coli-aerogenes group of gram-negative rods, which produce lactic acid among other substances; also coloured micrococci, and some very small gram-positive rods, perhaps related to the propionic bacteria, which occur sometimes in faeces and were named by Orla-Jensen Microbacteria. These last are facultative anaerobes but contain catalase (Doetsch and Pelczar, 1948). Of the three known species, one is motile and thermophilic (optimum 50° C); it is homofermentative (Bolcato, 1957). Furthermore, recent work on removing bacteria from milk by centrifuging has led to the recognition that much of the flora consists of saprophytic Mycobacteria, some of which are thermoduric.

The terminology of Orla-Jensen has been altered by Bergey *et al.*, who have combined all the rod-shaped lactic bacteria into one genus, *Lactobacillus*. The word is unfortunate, but the grouping is logical and is now generally used. The cell walls of the rods are very similar both in amino acid and in sugar content (Ikawa and Snell, 1960). Only *L. casei* is separated from the others in containing little or no ribitol phosphate but has a glycerol-phosphate instead.

Their very weak ability to synthesise amino acids and vitamins, is another common and characteristic property of the group. Furthermore, because the rod-shaped and the coccoid forms have so much in common, Orla-Jensen combined them together with the propionic acid bacteria into one family the *Lactobacteriaceae* which is still recognised.

The Streptococci

Within the streptococci, classification is often made on the basis of "hemolysis," i.e., discolouration surrounding the colony on blood-agar plates. Those producing a green zone, without lysis of the erythrocytes (a-hemolysis), are called the "viridans" group, and include *Sc. thermophilus, Sc. lactis, Sc. bovis, Pneumococcus,* and an organism called *Sc. salivarius,* found in the mouth and on restaurant glassware. *Sc. bovis,* which lives in the rumen of cattle, is exceptional in being a true anaerobe (Hungate *et al.*, 1952). The streptococci producing a colourless zone in which the erythrocytes are lysed are the truly haemolytic. These were subdivided by Lancefield (1925) on serological grounds as follows: Group A: *Sc. pyogenes,* including all the important *human* pathogens. Group B: *Sc. agalactiae* (= *Sc. mastiditis),* the cause of mastitis in the udder of the cow and thus a common form in milk. Group C: the so-called "animal pyogenes," causing many infections of animals, and *Sc. equi,* the cause of strangles in horses. Group D: *Sc. durans,* a thermoduric organism found in pasteurised milk, *Sc. zymogenes, Sc. liquefaciens,* which liquefies gelatin, and *Sc. fecalis,* from animal intestines. The last includes a motile strain. This group D has been termed the *Enterococci* (Sherman, 1938). Skadhuge (1950) considers it as having only three true species, *Sc. durans, Sc. fecalis,* and *Sc. pseudofecalis.* Groups A and C produce a characteristic viscous polysaccharide different from those of Pneumococci, and probably identical with hyaluronic acid (Lancefield, 1954).

This substance is the intercellular material of animal tissues; it is a polymer of N-acetyl-glucosamine and glucuronic acid. The Pneumococci are near to the viridans group but are distinguished by greater fragility of their cell walls. For example, they are dissolved by bile or 10 per cent bile salt (sodium taurocholate). When treated with penicillin, horse serum, and a Mg salt in presence of 10 to 20 per cent sucrose, they readily give rise to "L-forms"; these grow into colonies with large bodies (possibly fused wall-less cells) and long chains of fine granules, quite unlike the L-forms of streptococci (Madoff and Dienes, 1958). For the same reason pneumococci autolyse rapidly, so that old cultures often lose their grampositivity. A few strains are strict anaerobes though otherwise typical; these can mutate to aerobic forms (Smith, 1936).

Both Streptococci and Lactobacilli occur in moderate numbers in the human mouth and may be the cause of dental caries, through the acid they produce. Four homofermentative and three heterofermentative species of rods, and four Streptococcus species are found. Other common mouth organisms include Leptotrichialong, slightly tapered rods or filaments and two spirochaetes of the genus Treponema.

Lactic Fermentation

Products: The fermentation properly can be very nearly expressed by the simple equation:

$$C_6H_{12}O_6 \longrightarrow 2\ C_3H_6O_3$$

In animal tissues this reaction is termed *glycolysis* (literally, "sugarsplitting," although the splitting of the hexose into trioses is only one of 11 reactions which together constitute the lactic fermentation). The homofermentative forms, especially the Thermobacteria, give nearly quantitative yields of lactic acid from hexose sugars in acid or unbuffered media. Alkaline media favour the formation of other products; for instance, *Sc. liquefaciens,* which at pH, 5 gives 87 per cent lactic acid, yields only 61 per cent lactate at *pK* 9 (Gunsalus and Niven, 1942). Along with lactic acid are produced small amounts of formate, acetate, and alcohol, approximately in the ratio 2:1:1 (Friedemann, 1939).

The optical rotation of the lactic acid is characteristic of the organism. A bacterium that forms D-lactic acid in milk will form D- lactic acid in any other medium "whether the source of energy be alcohols, aldoses, ketoses, pentoses, hexoses, or polysaccharides" (Orla- Jensen). The same is true for organisms producing the L- or i-forms.

Thus the optical activity does not derive directly from that of the sugar. However, the optically active hydroxy fatty acids formed by L. *casei* (Camien *et al.,* 1959) are probably formed directly from amino acids, like the fossil oils.

Some lactic bacteria show an induction period between inoculation and the onset of rapid fermentation (Fromageot and Roux, 1931, 1933). It is independent of sugar concentration, pH, but is shortened by adding phosphate; the shorter the induction period, the faster the sugar is subsequently fermented. This period is perhaps due to the formation of induced enzymes.

The initial fermentation rates vary with the organism, fructose being the most rapidly attacked by some, lactose by others.

Mechanism of the Homofermentative Process

The co-enzymes of yeast, DPN (NAD) and ATP, function equally well for the lactic fermentation of muscle, and indeed researches on yeast and on muscle have been carried out in parallel. The major difference between the alcoholic and lactic fermentations, and therefore between yeast and most of the lactic bacteria, is the absence of functional *carboxylase* in the latter. Apart from this the fermentations go through the same intermediate stages. When pyruvic acid is formed, therefore, it becomes the hydrogen acceptor for reduced DPN (in place of acetaldehyde in yeast) and is thus reduced directly, without decarboxylation, to lactic acid.

The protein part or apoenzyme of the hydrogen-accepting enzyme system is here lactic dehydrogenase, instead of alcohol dehydrogenase. To reduce pyruvic acid, the oxidation potential has to go lower than for acetaldehyde. Yeast contains lactic dehydrogenase too, but normally forms only traces of lactic acid, because it decarboxylates the pyruvic acid first. Lactic dehydrogenase is readily purified from yeast; it is a pink complex containing cytochrome b_2 and riboflavin phosphate linked to a colourless protein. The high yields of lactic acid from lactose mean, of course, that the galactose moiety of lactose is fermented just as well as the glucose. Lactose is readily hydrolysed to glucose and galactose by f-galactoidase, and the subsequent changes were elucidated by L.eloir and co-workers using strains of yeast that ferment galactose.

These organisms interconvert glucose arid galactose, using a special co-enzyme. uridino-diphosphoglucose or UDPG: When this is incubated with the enzyme, 25 per cent of the glucose in the molecule is converted to galactose, by inversion at the 4 position, a "Walden inversion" (discovered with chlorosuccinic acid by P. Walden in 1893). The enzyme is hence named galactowaldenase. The reaction occurs in two steps; first galactose-1- phosphate displaces glucose-1-phosphate in the UPPG and then the inversion takes place.

The UDPG is readily hydrolysed to release glucose-1-phosphate, and this is then converted to the fermentable 6-phosphate by phosphoglucomutase. In reverse these processes cause the synthesis of lactose in mammary glands, UDP-Gal plus glucose forming lactose rapidly (Watkins and Hassid, 1961). A whole family of similar systems is now known, interconverting several sugars, especially hexoses with the methylpentoses (Kornfeld and Glaser, 1960). The action of poisons has helped to elucidate the intermediate steps in this fermentation, though not primarily with the lactic bacteria. *Cyanide,* of course, has no effect, either on the fermentation or on oxidation by lactic bacteria, *Iodoacetate* prevents lactic acid formation and the discovery by Lundsgaard (1930) that muscle poisoned with iodoacetate could still contract disposed of the view that lactic acid formation is an essential result of muscular contraction.

The principal action of iodoacetate is exerted on the dehydrogenase of phosphoglyceraldehyde, whose SH group combines directly with the aldehyde *fluoride* poisons the conversion of phosphoglyeerie acid to Dhosphopyruvic acid (enolase reaction) by combining with Mg as a fluorophosphate-protein complex (Warburg and Christian, 1942) It also poisons phosphatases and phosphoglucomutase.

Formation of Lactic Acid from Pentoses and the Hetero-fermentative Mechanism

The mechanism, which is known as that of Embden, Meyerhof, and Parnas owes its characteristic high yields of lactic acid to the actin of aldolase which cleaves the hexose diphosphate into two parts, both of which form pyruvate and hence lactate. Also, two ATP are formed. With pentoses, which cannot be equally cleaved, there are two possible results: (1) the 2-carbon fragment appears as ethanol or acetic acid (or both), giving a mixture of products exactly like that produced by heterofermentative organisms from hexoses; or (2) the 2-carbon fragment recombines with a pentose molecule, and after a complex series of changes, the products are trioses.

Pyruvic Acid Oxidation and other Associated Reactions

The acetate and CO_2 that appear along with lactate can also be formed aerobically from pyruvate by oxidative decarboxylation. *Sc. fecalis, M. aureus,* Gonococcus, and the fungus *Neurospora crassa* carry out the same reaction. Like the "plain" decarboxylation by pyruvate, it requires cocarboxylase (diphosphothiamine); thus the lactic organisms, not possessing the apoenzyme of carboxylase, may still possess the co-enzyme. Fungi like Neurospora, which ferment pyruvate to alcohol, presumably have both. A Neurospora mutant which could not oxidise pyruvate could still make alcohol and hence is probably lacking only the pyruvic oxidation factor (Strauss, 1952).

The oxidation of pyruvate is a complex process and can go in more than one way. Lipmann (1940), using an *enzyme* prepared from *L. delbruckii,* found that phosphate was essential and identified an intermediate compound as acetylphosphate, $CH_3COOH_2PO_8$. However, at least with *Sc. fecalis,* this is not the first but the fourth step in the series of reactions.

Cell-free enzymes prepared from *Sc. fecalis* oxidise pyruvate only in the presence of a special "oxidation factor" (Gunsalus *et al.* 1952), which in *E. coli* is a complex of cocarboxylase and the amide of dithio-octanoic acid, called lipoic acid (Reed and de Busk, 1952-1953; Reed, 1953). In its absence the extracts oxidise glucose only to pyruvate, which may accumulate as such or be condensed to acetoin.

The first step in pyruvate oxidation is the same as in its anaerobic decarboxylation; the pyruvate combines with thiamine pyrophosphate (TPP), most probably on the 2-carbon of the thiazole ring, (Breslow, 1958; Breslow and McNelis, 1959). Pyruvate combine with this intermediate, and when it does so it tends to withdraw electrons from the carboxyl group and release the CO_2. Thiamine alone, in solution, catalyses detectably the decarboxylation of pyruvate and other reactions for which cocarboxylase is co-enzyme, including the condensation to acetoin (Mizuhara *et al.*, 1951).

The reduced lipoic acid that is liberated is dehydrogenated by DPN and thus reconverted to its original S-S form, and the DPNH passes on the hydrogen to a flavoprotein, whose reoxidation by air (in the case of *L. delbruckii)* produces H_2O_2. Note that this is the *only* oxidation step in the whole process; it is here that the per cent O_2 of equation (1) participates. In the fourth step, acetyl-co-enzyme A (in presence of the enzyme transacetylase) comes to equilibrium with acetyl phosphate.

$$CH_3CO{-}SCoA + HO{-}H_2PO_3 \rightarrow CH_3CO{-}O{-}H_2PO_3 + CoA{-}SH$$

Finally, since acetyl phosphate has a high-energy phosphate bond, it can phosphorylate ADP. In so doing it is converted to acetic acid:

$$CH_3COO{-}H_2PO_3 + ADP^{35} \rightarrow CH_3COO^- + ATP^- \quad (6)$$

The same reaction was noted above when xylulose-phosphate is cleaved by phosphoketolase to produce triose and acetyl phosphate. Alternatively, acetyl phosphate is directly hydrolysed, though slowly, to acetic acid and free phosphate. To a small extent pyruvate can be split into acetyl-CoA and formate, as by the formic bacteria. The hydrogen that would have reduced the pyruvate to lactate probably reduces the acetyl-CoA to ethanol, thus accounting for the products. Also, some organisms convert pyruvate to acetate anaerobically by using a second molecule of pyruvate as hydrogen acceptor.

L. brevis carries out reactions anaerobically, while in air the lactate is removed by oxidation to acetate and CO_2 (Walker, 1959). Soluble preparations from *E. coli* and *Sc. fecalis* carry out reactions, accumulating acetyl phosphate (Korkes *et al.*, 1952). Further, if the condensing enzyme is present and oxalacetate is added, these preparations form citrate (Stern *et al.*, 1952): 2 Pyruvate + Oxalacetate $\rightarrow$ Citrate + CO_2 + Lactate Evidence for the participation of lipoic acid is given by the powerful inhibition exerted by arsenite, which combines with its two adjacent S atoms. An alternative to lactic acid formation is the reaction in *L. delbruckii* and *Proteus uulgaris,* where pyruvic acid, linked with cocarboxylase, reduces FAD (Moyed and O'Kane, 1952; Hager *et al.,* 1954). In Proteus this is reoxidised via the cytochrome system. Arsenite does *not* inhibit this process, and indeed lipoic acid was shown to be absent from the purified enzyme. The products are acetyl phosphate and COO. Third, *Mycob. phlei* and *Mycob. avium* oxidise lactate to acetate and C(X without going through pyruvate (Sutton, 1954-1957). The enzyme, which contains riboflavin-5-phosphate, is in fact inhibited by pyruvate, and the intermediate appears to be a free radical (Commoner *et al.,* 1958). Since neither of these bacteria is a lactic organism, the relationship to the lactic system is probably remote.

Other Mechanisms of Lactic Acid Formation

Glycerol can be oxidised to lactic acid by *Sc. fecalis.* If catalase is added, the reaction goes more vigorously and only a little per cent of O_2 is used per mole of glycerol oxidised. This reaction apparently goes by way of phosphoglycerol, triosephosphate, and phosphoglyceric acid (Gunsalus and Umbreit, 1945). The formation of lactic acid by fungi of the genera Rhizopus and Mucor, especially *M. rouxii,* is accompanied by other acids and ethyl alcohol (Takahashi and Asai, 1933). These fungi do not attack lactose. Anaerobically the mycelium of *Rhizopus MX* produces almost equal quantities of lactic acid, alcohol, and CO_2, essentially as with *Ln. mesenteroides.* Aerobically, alcohol production is largely suppressed (Waksman and Foster, 1939); instead, carbon is assimilated into the mycelium and some fumaric acid is formed (Margulies and Vishniac, 1961). The yield of lactic acid is unchanged in air, and the earlier belief that it was increased is explained by formation of the fumaric acid. With C^{14} in the 2 position of glucose, the isotope is mainly found in the carbinol of lactic acid and of ethanol; when the C^4 is in the 1 position, it appears as the methyl group and also (in air) in the fumaric acid. In addition, some $C^{14}O_2$ is fixed into the fumaric acid carboxyls. Margulies and Vishniac conclude that the fermentation is essentially by the Embden-Meyerhof route (as with Leuconostoc), and indeed all the enzymes needed for this pathway are present in the Rhizopus.

The formation of fumarate is probably due to the combination of CO_2 with phosphoenolpyruvate to form oxalacetate, followed by its reduction to fumarate. *Rhizopus MX* can also form some lactic acid from formate, by a complex condensation with a 2-carbon unit (Jefferson *et al.*, 1952). C^{14} formate yields the C^{14} mainly in the methyl group of lactate. Yeasts also form very small amounts of lactic acid in the course of fermenting sugars to ethanol. While some of this may be derived from alanine or other amino acids, just as are the fossil oils, some lactic acid is formed from sugar (about 1 per cent yield; Hohl and Joslyn, 1941).

One of the Zygosaccharomyces, *Z. acidifaciens,* which produces acetic acid, alcohol, and glycerol anaerobically as in Neuberg's third form of fermentation, makes appreciable amounts of lactic acid (Nickerson and Carroll, 1945), although the maximum yield is only one-fifth to one-tenth of that produced by *Sc. lactis* growing on milk. Several lactic bacteria decarboxylate malate to lactate and CO_2. These organisms are important in wine making, since they can destroy completely the malic acid in grape juice. The strains isolated include a Leuconostoc and a Lactobacillus similar to *L. plantarum* (Radler, 1958). Apparently fruit juices contain all the necessary amino acids and vitamins for their growth. The water mould *Blastocladia pringsheimii* produces from glucose 85 per cent of lactate and 11 per cent of succinate (Cantino, 1949). Aeration does not alter this ratio, and the mechanism of the reaction is not clear. If CO_2 is bubbled through the medium, more succinate is formed at the expense of lactate doubtless by CO_2 fixation.

Geotrichum Candidum

After milk has soured and been kept for some days, its surface often becomes covered with a fungus, *Geotrichum candidum* (formerly *Oospora lactis*). This organism is probably an imperfect member of the *Geotrichum candidum*.

A. and *B.* Hyphal tips.

C. Older hyphae broken up into oidia. 100 X.

D. Older hyphae showing vacuolation.

E. Hyphal tip and subapical region showing relatively thick transverse walls 220 X.

Ascomycetes, related to Candida, although its position is not fully clear. Its dichotomously branching mycelium breaks up, especially on lateral branches, into rectangular or oval fragments called oidia, each of which can give rise to a new mycelium. Geotrichum is oxidative in its metabolism and destroys the lactic acid of sour milk. However, there are strains which grow on lactose rather than lactic acid (Geffers, 1937). Cultures on wort agar produce some acid and later destroy it again. It is also proteolytic and deaminates amino acids. For these reasons the *pH.* of soured milk may return to neutrality or even go alkaline when infected by it. Geotrichum forms quantities of fat, some strains containing as much as 50 per cent of the dry weight in the form of neutral fat (Geffers, 1937). Some commercial interest attaches to the fact that it can be grown on acidhydrolysed straw or oat-hulls and will form 22.5 per cent fat on such media. Unlike the lactic bacteria, Geotrichum possesses catalase and can synthesise riboflavin and other vitamins (Kuchar, 1950).

General Physiology of Lactic Bacteria

True lactic organisms do not produce large colonies on ordinary solid media. This is due to three peculiarities, whose understanding has required about 30 years:

1. They are microaerophilic for the reason that they possess no cytochrome oxidase or catalase and can respire only by spontaneous oxidation of reduced flavoprotein. This process produces H_2O_2, which thus can accumulate and inhibit growth.

Some heterofermentative rods, like *L. cereale,* are exceptions, while some streptococci have a flavoprotein peroxidase, which uses up the H_2O_2 to oxidise DPNH (Dolin, 1953-1955). A few streptococci, e.g., *Sc. bovis,* are true anaerobes.

2. Their ability to synthesise amino acids is small. Some peptones are inadequate for growth, because of low content of one or another amino acid. Casein peptone is generally reliable, but Orla-Jensen (1919) found yeast extract better for the Streptobacteria and Thermobacteria.

In recent years pure amino acids have been used as a medium, together with sugar, phosphate, Fe^{++} per cent Mg^{++}, Mu^{++} per cent acetate, and the bases, adenine, guanine, and uracil. Vitamins are needed in addition. The relative concentrations are often critical, because even low levels of cystine can be inhibiting, while high concentrations of alanine or leucine inhibit growth, by preventing the utilisation of glycine or isoleucine, respectively (Kihara and Snell, 1960). In the latter case, growth can be reinstated by large amounts of glycine or isoleucine, or, even better, by their peptides. Sometimes the use of peptides is preferable, e.g., peptides of tyrosine for *Sc. fecalis,* of histidine for *Pediococcus cereviseae* (Florsheim *et al.*, 1962), of both for *L delbriickii,* and of glycylserine for Leuconostoc, since the growth is then less subject to inhibition by related amino acids (O'Barr and Pierce, 1960).

Nevertheless, the nutrient requirements are very useful, since any one constituent of the medium can be omitted and an unknown preparation added in varying amounts; the extent of growth will then be a measure of the amount of the missing amino acid in the unknown. *L. arabinosus* strain 17-5, *L. casei, Leuconostoc mesenteroides* (strain P- 60) (= *Sc. equinus), L. delbriickii,* and *Sc. fecalis* have been mainly used. In these assays growth is determined either by the turbidity of the medium or simply by the concentration of lactic acid formed under stated conditions, such as 72 hours at 35°C. There is an extensive literature on this, mainly dealing, however, with the amino acid composition of proteins (cf., e.g., Neumann, 1949, and the review of Snell, 1945).

By contrast, the "pseudo-lactic" bacteria, the coliforms, grow well with single amino acids, or even ammonium, as source of nitrogen, and sugar or lactic acid as carbon source. The amino acid requirement makes the growth of lactic bacteria very susceptible to inhibition by amino acid analogues. For example, cyclohexyl-alanine competes with phenylalanine, and cyclopentyl-alanine with leucine (Pal *et al.*, 1956). *Ln. dextranicus* and *L. arabinosus* are powerfully inhibited by 3-aminomethyl-cyclo-hexylglycine, which competes with lysine.

3. The lactic bacteria also need vitamins of the B group. The need for riboflavin will be obvious because of their flavoprotein system, and the requirement is relatively large. Several lactic bacteria have, therefore, been used for the determination of riboflavin in unknown materials (Snell and Strong, 1939; Williams *et al.*, 1941). Nicotinic acid, thiamine, folic acid, p-aminobenzoic acid, pyridoxine (vitamin B_6), and vitamin B_{12} are also required by many strains and can therefore be assayed by omitting any one of them from the medium, and measuring the growth, or the lactic acid, when known amounts have been added. The use of the lactic organisms for bioassay of vitamins has become widespread in recent years, but it needs caution. For instance:

(a) The growth-promoting activities of pyridoxal and pyridoxamine, two forms of vitamin B_6, are very different.

(b) These B_6 vitamins are often more active in the

phosphorylated form, and indeed *L. lactis* cannot use them except when phosphorylated (Hendlin *et al.*, 1950).

(c) If *Sc. fecalis* is given B_6, it decarboxylates tyrosine in the medium and has to be supplied with tyrosine in the form of a peptide (Kihara *et al.*, 1952). Peptides, both of tyrosine and histidine, also stimulate *L. delbruckii.*

(d) Several different forms of pteroylglutamic acid (folic acid) exist, with widely different activities for different organisms, one (Leucovorin) being as much as 10^5 times as active as folic acid on *Ln. citrovorum* (Wieland *et al.*, 1952).

(e) *L. arabinosus* and L. *casei* can synthesise folic acid if *p-amino*benzoic acid (Sarett, 1947) or certain amino acids (Harry, 1951) are given.

(f) The response to vitamin B_{12} (by *L. leichmannii)* depends on the oxygen tension, so that repeatable results are only obtained if a reducing agent is present in the medium.

(g) Thymidine will substitute for B_{12} and also for folic acid (after a lag period) in the case of *L. leichmannii* (Franklin *et al.*, 1949). (h) *Ln. citrovorum* 8081 does not require either folic acid or thymidine if it is supplied with a purine-like factor present in liver (Sauberlich and Baumann, 1948, 1949).

(h) *Sc. fecalis* and some Lactobacilli do not require any aspartic acid if biotin is present (Stokes *et al.*, 1947).

(i) *L. casei* in the absence of vitamin B_{12} requires D-alanine, as well as L-alanine which must be supplied as a peptide (Kihara *et al.*, 1952).

(j) Biotin in yeast is present as biotinyl-lysine (Wright *et al.*, 1952).

Points *a, b,* and *d* make the vitamin assays difficult to interpret; points have the result that the vitamin and amino acid assays can interlock. No doubt some of the differences between different peptones as substrates for lactic bacteria, recorded in the older literature, were due to different contents of vitamins rather than of amino acids. Besides amino acids and vitamins, several purines and pyrimidines, and one or more fatty acids may be required. Lactobacillic acid, a cyclopropane derivative of stearic acid related from *L. arabinosus* and *L. casei* and also from *Agrob. tumefaciens,* in which it comprises 13 per cent of the cell lipids. This and three related acids promote the growth of *L. arabinosus, L. casei,* and *L. delbruckii* on biotin-deficient media; they may be essential metabolites (Hofmann and Panos, 1954).

The great spread in concentrations required, between nutrients and salts at one end and vitamins at the other, is notable. The extremes, dextrose and biotin, show a ratio of 50 million times in concentration. It is clear that with the appropriate omission such a medium can be used to assay almost any amino acid or vitamin, and the Lactobacilli have become laboratory reagents.

Propionic Bacteria

The Propionic bacteria are closely related to the lactic bacteria. They are gram-positive, non-sporeforming, non-motile rods. Aerobic cultures show rods somewhat on the large side (as bacteria go) but "mostly bent, locally swollen, and sometimes branched" (van Niel, 1928;). These general properties suggest relationship to the Corynebacteria.

However, their physiology does not resemble that of most Corynebacteria, which are typically aerobic. Only in two cases are there indications of a relationship. The pathogen *C. diphtheriae* does convert sugars to acids under partially anaerobic conditions (de Wolff, 1927), and although this is not a typical propionic fermentation, still some of its strains form considerable amounts of propionic acid, along with acetic, lactic, and succinic acids (Tasman and Brandwijk, 1938).

Another Corynebacterium, C. *acnes,* is actually

microaerophilic and does carry out a propionic fermentation; on this account some would class it as a Propionibacterium (Douglas and Gunter, 1946). Both these organisms are morphologically of the diphtheroid type. The true Propionibacteria grow under anaerobic or at least microaerophilic conditions, like the homofermentative Lactobacilli. Unlike these, however, they contain a cytochrome system (Chaix and Fromageot, 1942), and generally catalase also (Virtanen, 1931; Hitchner, 1934). However, they never produce large colonies on the surface of solid media. They can oxidise glycerol and various organic acids, probably via the tricarboxylic acid cycle (Delwiche and Carson, 1953).

Eleven species of Propionibacteria have been described. On yeast extract-glucose media two *(Pr. freudenreichii* and *Pr. shermanii)* grow as small streptococci, seven (including *Pr. jensenii)* as rods, and two (Pr. *arabinosum* and *Pr. pentosaceum)* as highly irregular long rods or spheres. Further distinction is by their carbohydrate fermentations and their pigments.

Pure cultures were first obtained using enrichments inoculated from Swiss cheese, by von Freudenreich and Orla-Jensen (1906). This is a hard cheese with large holes in it, produced by gas bubbles. The propionic acid is partly responsible for its characteristic taste and smell, and it contains 10,000 to 300,000 propionic bacteria per gram.

Other Organisms Forming Propionic Acid

Besides the Propionibacteria, several other bacteria produce propionic acid. An important group exists in the rumen of cattle and other animals; some of these are anaerobic gram-negative cocci (e.g. *Veillonella gazogenes = Micrococcus lactilyticus,* and *M. aerogenes),* while others are gram-positive and resemble the *C. acnes;* these readily form propionic from lactic acid (Gutierrez, 1953).

The micrococci can produce propionic acid from pyruvic, α -ketoglutaric, and lactic acids (Whiteley and Ordal, 1957), as well as from sugar. However, lactic acid is not an important intermediate in the rumen itself (being mainly converted to acetate there), nor do the rumen bacteria attack it very rapidly (Jayasuriya and Hungate, 1959). Propionic acid can also be formed from alanine, lactic and pyruvic acids, and other compounds by *CL. propionicum,* although by a different route from that in the Propionibacteria. Finally, *Rhodo-sp. rubrum,* acting on pyruvate in the dark, produces propionic and acetic acids (Kohlmiller and Gest, 1951).

Fermentation

Formation and Dehydrogenation of Pyruvic Acid: Propionic acid is formed by the Propionibacteria from glucose, lactose, or, with some species (e.g., Pr *arabinosum),* from pemoses. It can also be formed from lactic acid, malic acid, and glycerol The early stages in hexose fermentation by propionic bacteria are almost certainly the same as with yeast; the dried bacteria do form hexosephosphate (Virtanen and Karstrom, 1931). They can transfer phosphate groups from ATP to glucose, glycerol, arabinose, and erythritol; the phosphoglycerol, in presence of fumarate as hydrogen acceptor, is dehydrogenated to phosphoglyceric acid (Barker and Lipmann, 1949). Toluene-treated cells can also form phosphoglycerate from hexosediphosphate, or from hexose, using pyruvate as hydrogen acceptor (Werkman *et al.,* 1937). These reactions have to be carried out in presence of fluoride, in order to inhibit hydrolysis of the esters by phosphatases. It can be concluded that the fermentation uses essentially intermediates, and involves the same phosphorylations, as the alcoholic fermentation in yeast and the glylolytic process in animal tissues. Metabolism of pentoses probably involves the interconversions for lactic bacteria. In the case of arabinose, which is fermented by *Pr. arabinosum* and *Pr. pentosaceum,* a highly specific enzyme isomerising 5-phospho-D-arabinose to 5-phospho-D-ribulose has been demonstrated (Volk, 1960). No doubt this is subsequently converted to the fermentable 5-phospho-D-xylulose. In cheese the main substrate is lactic acid, and it was this fermentation that was first studied.

Von Freudenreich and Orla-Jensen (1906) found with pure cultures, however, that the ratio *propionate: acetate* was not quite 2, but 1.6 to 1.8; van Niel (1928) generally obtained similar results, but found the ratio *acetate:* CO_2 to be always 1. The constancy of the latter ratio suggested that the acetic acid and CO_2 must arise from the same precursor, which would have to be pyruvate. The reaction must, therefore, be similar to the oxidative decarboxylation of pyruvate but without oxygen:

$$CH_3COCOOH + H_2O \longrightarrow CH_3COOH + CO_2 + 2(H)$$

Instead of 3 moles of lactic acid, moles of glucose could be substituted: the equation then describes very well the results in the literature.

When pyruvate is dehydrogenated by *M. aureus, Gonococcus* and *Sc. fecalis* under anaerobic conditions, a second molecule acts as hydrogen acceptor and is reduced to lactate (Krebs, 1937):

$$2\ CH_3COCOOH \longrightarrow CH_3COOH + CO_2 + CH_3CHOHCOOH$$

This "dismutation" is suggestively similar to one of the two main routes whereby bacteria form propionic acid. It is in the hydrogen acceptors that the two kinds of propionic fermentation differ.

Formation of Propionic Acid by CL. propionicum

It will be convenient to consider this special case first. *Cl. propionicum* ferments pyruvate and lactate with the same overall products as the Propionibacteria and in addition ferments alanine, serine, and acrylic acid as follows (Cardon and Barker, 1947):

$$3\ CH_3CHNH_2COOH + 2\ H_2O \longrightarrow$$
$$2\ CH_3CH_2COOH + CH_3COOH + CO_2 + 3\ NH_3$$

$$3\ HOCH_2CHNH_2COOH + H_2O \longrightarrow$$
$$CH_3CH_2COOH + 2\ CH_3COOH + 2\ CO_2 + 3\ NH_3$$

$$3\ CH_3CHCOOH + 2\ H_2O \longrightarrow$$
$$2\ CH_3CH_2COOH + CH_3COOH + CO_2$$

Each reaction can evidently be considered as an oxidation of one or two molecules of the substrate and reduction of the remainder to propionic acid. The intermediate, and direct precursor of propionic acid, would be acrylic acid. The correctness of this interpretation is shown by what happens with threonine (Cardon and Barker, 1947). This amino acid yields a mixture of butyric and propionic acids.

Thus, two molecules are deaminated, decarboxylated, and oxidised to propionic acid; this provides 4(H) with which the third molecule is hydrogenated to butyric acid. Broof that in these organisms propionic acid arises by direct reduction of a 3-carbon precursor is given by labelling experiments (Leaver *et al.* 1955). Lactate with C^{14} in the C-3 (methyl) position gave rise propionate labelled *only* in the C-3 position.

By contrast, with Propionibacteria *both* the C-3 and the C-2 positions of the propionate contained isotope. The same was true using pyruvate lactate labelled in the 2-position. Evidently, therefore, the C-3 and C-2 carbons of propionate *become equivalent* at some stage in the process, and this can only mean that succinate is an intermediate. Indeed, the addition of a second carboxyl group to pyruvate (as first step on the route to succinate) one of the two "key reactions" of the propionic fermentation.

Steps in the Propionic Fermentation Proper

It was originally thought that the 4 (H) produced in the conversion of lactate to pyruvate and of pyruvate to acetate were directly transferred by Propionibacteria to lactate, reducing it to propionate, as with *Cl. propionicum:*

$$2\ CH_3CHOHCOOH + 4(H) \longrightarrow 2\ CH_3CH_2COOH + 2H_2O$$

An early indication of this was Barker and Lipmann's finding (1944) that dried bacteria treated with fluoride can make propionate from pyruvate but not from lactate. The only other simple intermediate between pyruvate and propionate is very late; yet acrylate yielded no propionate. Since several

Propionibacteria could form propionate by decarboxylating succinate (Delwiche, 1948; Johns, 1951), a roundabout route from pyruvate via the 4-carbon acids was possible.

Succinyl-co-enzyme A appeared to be an intermediate (Whiteley, 1953; Wood *et al.*, 1956). Succinate was found to be decarboxylated, too, by bacteria growing in the rumen of cattle, both the anaerobic cocci alone (Johns, 1951; Whiteley, 1953) and preparations of total rumen bacteria. Since some other rumen organisms *accumulate* succinate, it is apparent that they fail to decarboxylate it, and thus all these organisms probably operate by pathways similar to those of the Propionibacteria. The total rumen organisms actually form propionate from succinate more rapidly than do the Propionibacteria.

At first there were conflicts and inconsistencies in regard to succinate decarboxylation; some Propionibacteria seemed not to attack succinate, and labelling experiments with $C^{14}O_2$ and other compounds did not agree well with the proposed route. Fortunately these have now been resolved through experiments on the process going in the opposite direction, namely, the *oxidation* of propionate by the mitochondria of animal tissues (Flavin *et al.*, 1955; Lardy and Adler, 1956). These results show that in the oxidation of propionate to succinate:

(1) CO_2 is taken up,
(2) Both ATP and co-enzyme A are needed, and
(3) Some tissues produce not only succinate but also its isomer, methyl malonate, HOOC-CH-COOH+CH_3.

Ochoa and his co-workers speedily explained this last fact by showing that methylmalonic and succinic acids, in the form of their CoA derivatives, are equilibrated by an isomerase enzyme in liver tissue.

The enzyme that carboxylates propenyl-CoA contains biotin. The application of these facts to the understanding of the propionic fermentation is direct. In the first place the organisms have long been known to form succinic acid, and in fact it was with them that the carboxylation of pyruvic to oxalacetic acid was discovered (Wood and Werkman, 1936, 1938). The succinic acid produced here, and in the alcoholic fermentation, had been ascribed to the deamination of aspartic acid, but Wood and Werkman showed that the CO_2 evolved is actually taken up again in the propionic fermentation and an equal amount of succinic acid formed. It was deduced that oxalacetic acid was the intermediate.

This enzyme and a related one from *CL. tetanomorphum,* which isomerises glutamio acid to 0-methyl-aspartic acid, contain members of the vitamin Biz group as co-enzymes. The methylmalonyl-CoA isomerase uses dimethylbenzimidazolyl-cobamide (Stadtman *et al.*, 1960); it functions also in the flagellate Ochromonas, which cannot oxidise propionate when deficient in B_{12} (Arnstein and White, 1962). Both isomerases have been purified and studied in detail (Barker *et al.*, 1960-1961; Stjernholm and Wood, 1961). Because of these enzymes the Propiouibacteria are unusually rich in BU. When glycerol was the substrate, there were actually a *net* uptake of CO_2 by Pr. *pentosaceum* and an equivalent production of succinate (Wood and Werkman, 1940). Whereas on lactate 32 *fA* of CO_2 were evolved, on glycerol 355/J, CO_2 were *absorbed.* The CO_2 uptake was inhibited by fluoride, and at the same time formation of succinate was prevented. The reaction was confirmed with $C^{14}O_2$, because then the isotope was found in the carboxyl groups of *both* succinic and propionic acids (Wood *et al.*, 1940; Carson *et al.*, 1941). In the second place, participation of co-enzyme A in the fermentation was indicated in several ways; the cell-free enzyme from *M. lactilyticus* requires ATP and CoA to convert succinate to propionate (Whiteley, 1953), and, in the reverse process, extracts of *Pr. pentosaceum* carboxylate propionate to succinate rapidly in presence of CoA but only slowly in its absence (Phares *et al.*, 1956). A purified extract from *Pr. shermanii* will transfer carboxyl from succinate to propionate only if CoA is added (Swicfc and Wood, 1960).

In the third place, the *Pr. shermanii* extract will rapidly decarboxylate methylmalonyl-CoA to propionate, while succinyl-CoA is only one-tenth as effective (Swick and Wood, 1960). The reaction goes in presence of pyruvate, which accepts the carboxyl and becomes oxalacetate.

No Acid other than Pyruvate will Accept the Carboxyl: Fourth, the enzyme system transfers co-enzyme A from propenyl CoA to succinate, forming succinyl-CoA and thus completing the cycle. A complication is that while the intact bacteria certainly fix CO_2 in the process, the purified extracts do not. When the methylmalonyl- CoA carboxylates pyruvate, the carboxyl does not come off as CO_2 but remains attached to the enzyme. This enzyme is thus a *transcarboxylase;* it contains biotin as co-enzyme and is strongly inhibited by the biotin-combining compound avidin (Swick and Wood, 1960; Lynen, 1961).

The carboxyl transfer explains some of the discrepancies alluded to above, especially in regard to the uptake and transfer of $C^{14}O_2$. (The other discrepancies are probably due to failure of some preparations to form succinyl-CoA.) Both the transcarboxylase and also the isomerase that interconverts methylmalonyl-CoA and succinyl-CoA have recently been purified from Pr. *shermanii* (Wood and Stjernholm, 1961; Stjernholm and Wood, 1961).

It is a wry twist that the very reaction in which CO_2 fixation was discovered should turn out not to involve CO_2 fixation as an essential part. However, when sugar is fermented by intact cells, some at least of the CO_2 is released and reabsorbed, since CO_2 comes off, and isotopic COo enters the carboxyl groups. Besides, the yield of propionate from succinate with *M. lactilyticus,* and of propionate from pyruvate with Rhodospirillum, is strongly influenced by the CO_2 pressure (Johns, 1951; Kohlmiller and Gest, 1953). The fact that some of the rumen bacteria produce succinate and others propionate also indicates that the two acids are alternative final products, and therefore that transcarboxylation can be replaced by net CO_2 uptake in some cases. When CO_2 is taken up, the most probable pathway is, of course, the carboxylation of phosphoenol pyruvate to oxalacetate.

Direct carboxylation of pyruvate to oxalacetate does not readily occur because the equilibrium lies in the direction of decarboxylation; energy can be supplied to drive it either by converting the high-energy phosphate bond to inorganic phosphate, as here, or else by simultaneously hydrogenating the product to malate, as by the "malic enzyme of green plants, which converts pyruvate directly to malate. A suggestive parallel to the formation of propionate is furnished by the bacterial decarboxylation of malonate to acetate and CO_2 (Hayaishi, 1953).

A purified enzyme from the aerobic *Ps. fluorescens* brings this about, but only in presence of ATP and CoA, so that malonyl-CoA is evidently an essential intermediate. It can be appreciated, too (since COOH is an electron-withdrawing group), that the two carboxyls linked to the same C-atom in malonate and methylmalonate would be much more reactive than those of succinate.

Integration

Finally, the hydrogen balance of the whole fermentation should be noted. Conversion of triose to pyruvate yields 2H. When the pyruvate has been carboxylated to oxalacetate, 4H are required to reduce it to succinate. Formation of *two* moles of pyruvate can thus produce only *one* of succinate (hence, one of propionate). What happens to the second pyruvate. The answer is that a third mole of triose yields pyruvate and this pyruvate is converted to acetate, CO_2 + 2H. The resulting 4H + CO_2 gives rise to a second succinate. Both the CO_2 and the CoA cycle continuously during the fermentation, so that it is represented *in toto*.

Formic Fermentation: The production of formic acid is the common result of *a* group of related fermentations with many variants. These variants have so confused the issue that the simple term *formic fermentation* has not been used. However, this term characterises the process very well, since its "key-reaction," is:

$$H_2O + CH_3COCOOH \longrightarrow (CH_3COOH) + HCOOH$$

The acetic acid is bracketed because it is not an essential product, while the formic acid is.

Enterobacteriaceae and their Significance: The formic-fermenting bacteria, often termed the coli-typhosum or colon-typhoid-dysentery group. One of them, *Tscherichia coli* or *Bacterium coli* is the most widely used of all bacteria both in biochemical and genetical researches. The name *Enterobacteriaceae* was put forward by Rahn (1937) because so many of them are inhabitants of the animal intestine. Earlier they were sometimes called the enteric bacilli, but the name *Bacillus* has long been restricted to sporeformers by all but the most inveterate medical writers.

These bacteria are gram-negative rods, with their flagella always peritrichous, if present at all. They all produce acid from glucose, and most of them from other carbohydrates as well. Unlike the lactic bacteria, however, they are facultatively aerobic and contain catalase. Many strains of *E. coli* will grow well on ammonium lactate with KH_2PO_4 and salts;

Aerobacter will grow even with citrate as the sole carbon source. Their morphology is simple, except for *Proteus,* which is claimed to form pleomorphic rods in young cultures; however, these are doubtless really chains of short rods of which cross walls are not easily seen. Some species of Proteus spread rapidly over agar in a characteristic swarming movement. Fimbriae are very common in the group. These thin, stiff filaments are found on virtually all strains of Proteus and Serratia, and on 60 to 80 per cent of Salmonella and Klebsiella as well as on other genera (Duguid and Gillies, 1957-1958).

Some members of the group, .particularly Proteus and Aerobacter, are normally present in soil, but *E. coli* is a regular inhabitant of the large intestines, and therefore of the faeces of mammals. It grows simply on the nutrients of the faeces and is therefore not a parasite, but rather a *commensal* with the host. Researches on vitamin K, however, have cast a new light on so-called commensalism. Mammals and birds need vitamin K for the normal clotting of blood; haemorrhages occur in its absence. Nevertheless, all but the newborn are independent of the vitamin in the diet, because it is synthesised by bacteria in the intestines. Pure cultures of *E. coli* on a synthetic medium synthesise and excrete into the medium relatively large amounts of vitamin K-active substances (Dam *et al.*, 1941). We should therefore regard *E. coli* as a symbiont rather than a commensal. The formation of vitamin B_{12} from cobalt by rumen bacteria and probably by *E. coli* (Barker *et al.* 1960) is a comparable case.

The synthesis of vitamins is not limited to vitamins K and B_{12} nor, indeed, to the activity of *E. coli.* Other members of the *Enterobacteriaceae* do the same, and so do some of the aerobic sporeformers. The amounts of B vitamins produced by some pure cultures on a strictly synthetic medium.

These are large. A yield of 115 micrograms of thiamine per gram of *E. coli* is interesting in regard to human nutrition. The human colon has a volume of about 4.5 litres, and some 40 per cent of human faeces represents bacteria, mainly *E. coli.* The colon could thus contain 1800 gm of bacteria which would correspond to 200 mg of thiamine. When it is considered that the daily human requirement is only about 2 mg, it is evident that the process of defecation, however desirable for other reasons, is unfortunate from the viewpoint of vitamin nutrition.

The presence of *E. coli* in such enormous numbers in faeces means that when it is found in water, milk, or food, these are probably contaminated with traces of faeces or of sewage. The identification of *E. coli* is, therefore, an important part of the routine bacteriological examination of foods and water. For this purpose distinction must be made between (1) intestinal types, including *E. coli* and pathogens, and (2) soil forms like Aerobacter. A number of laboratory tests have been worked out for this and are widely used by public health bacteriologists. These tests depend on differences in the fermentations.

The group as a whole may be subgrouped physiologically by their ability to produce hydrogen from dextrose and lactose, along with proteolysis as measured by gelatin liquefaction. It shows that the ability to attack lactose is typical of Escherichia and Aerobacter. A lactose-negative Escherichia ("E. *coli mutabile")* continuously throws mutants, about 1 in 100,000, which can attack lactose. *Aerobacter cloacae* is the species that liquefies gelatin. Serratia, or *B. prodigiosum,* is really to be considered as a group of red-pigmented *Aerobacter cloacae,* or in some isolates which do not attack lactose, as red-pigmented Proteus. All its isolates tested form acetoin, which is responsible for the "Voges-Proskauer reaction". The pigment is a pyrrole derivative and is apparently attached to the cell membrane (Purkayastha and Williams, 1960; Yoshida, 1962). Several organisms differing from Escherichia and Aerobacter only in that their fermentation of lactose is delayed have been grouped in a separate genus, Paracolobactrum (Bergey, 1957). But to give *generic* rank to what may prove to be only a single gene difference is quite unjustified.

A number of related organisms cause lesions, wilts, or soft rot in plant tissues. These bacteria, of which *B. carotovorum* is the best known, are common on market vegetables, on which they cause soft brown spots or areas, and some of them cause rot diseases on growing plants. In Bergey's manual they are treated as a separate tribe with the single genus *Erwinia.*

However, they clearly cause a formic fermentation, in most cases attacking lactose; in some acetoin is produced, which is characteristic of Aerobacter, while some are proteolytic. Frequently they will grow on citrate as sole carbon source, a behaviour typical of Aerobacter (Taylor, 1951). Thus they are closely related to Aerobacter. Their only distinguishing character is that of rapidly macerating plant tissue, especially carrot and turnip, which is not shown by other members of the *Enterobacteriaceae* (Elrod, 1942); nevertheless, this character alone scarcely justifies classifying them in a separate tribe. The same considerations hold for an animal pathogen.

"Friedlander's bacterium," causing some pneumonias in man. This differs from Aerobacter only in being heavily capsulated, but rough forms (R-forms) without capsule certainly exist, while Aerobacter often forms capsules too. Its listing in Bergey's manual as a separate genus, Klebsiella, therefore has little basis. Some taxonomists, indeed (cf. Wood, 1950), would not only include this organism in Aerobacter, but would also, and with good reason, eliminate Serratia as a separate genus. The Salmonella group is a rather homogeneous one. All are intestinal pathogens, causing enteric fevers and "food poisoning" of varying degrees of severity. Over 150 strains or isolates exist, distinguished mainly on serological grounds. The transduction phenomenon indicates a close interrelationship, and some workers would reduce them all to three species. *S. typhimurium (B. aertrycke)* is the commonest cause of food poisoning. The Salmonella fermentations are marked by failure to attack lactose, and, in the case of *S. lyphosa,* failure to decompose formate. They are not proteolylic. As a rule they can reduce trimothylamine oxide to trimethylamine, a reaction responsible for the characteristic smell of decaying fish.

The dysentery bacteria (genus Shigella) are less homogeneous. Their pathogenicity varies from slight to extreme, their serology is complex, and the fermentations vary, a few producing hydrogen and some attacking lactose slowly. Some reduce trimethylamine oxide; none is proteolytic. The genus as such may not have much more validity than Klebsiella. The intestinal diseases caused by these last two groups are characterised by inflammation of the intestinal wall. Because the area of tissue involved is quite large and relatively uniform, the onset of the diseases is typically very rapid. The flow of fluid to the area, which is characteristic of the "inflammatory process," leads to dilution of the faeces with fluid (as well as blood) and often thence to dehydration of the body as a whole. Cholera, caused by a vibrio, has many similarities to typhoid and dysentery. The endotoxins of these bacteria add a complicating factor to the disease. Of course normal strains of

E. coli do not attack the intestines in which they live, although strains biochemically indistinguishable from them may sometimes cause inflammations in the bladder and elsewhere.

"Coli" Type of Fermentation

Products: The products of the fermentation were first determined by Harden (1901).

Succinic acid was later added to the list of products. Later analyses give results agreeing in principle with Harden's, although not in detail. Two independent sets of analyses are summarised in table below:

Table: Products of E. coli Fermentations

(Experiments 1 and 2 in presence of $CaCO_3$ [Scheffer, 1928]; Experiment 3 held at *pR* 6.0 by adding ammonia [Blackwood *et al.*, 1956].)

Product	*Moles Per 100 Moles of Glucose Fermented Experiment*				
1	Experiment				
2	Ratio	Experiment			
3	Ratio				
CO_2	45	43	1	88	2
H_2	44	41	1	75	2
HCOOH	2	2	?	2	?
CH_3COOH	45	43	1	37	1
C_2H_5OH	45	40	1	50	1
Lactic acid	84	85	2	80	2
Succinic acid	27	31	W	11	11

Different members of the *Enterobacteriaceae* give significantly different products. Attention to these variations has been partly responsible for elucidation of the reaction mechanism. For example, *S. typhosa* gives little or no hydrogen gas, and in this case formic acid is found instead (Harden, 1901). With *Aerobacter aerogenes* the modifications are more extensive.

Many substrates besides sugars are fermentable. Glycerol gives more alcohol and less lactic acid. Ethylene glycol gives mainly lactic and acetic acids, while malic and tartaric acids yield acetic, succinic, and CO_2 in about equal amounts (Grey, 1924). Succinic acid itself is converted to formic and acetic acids and hydrogen. Glyceric acid yields SO per cent of acetic acid, formic acid, and a trace of alcohol; the L-form is used much faster than the D- (Virtanen and Peltola, 1930). Access of oxygen during the fermentation decreases the yield of formic acid and increases the lactic and succinic acids (Grey and Young, 1921). Acid *pR* has similar effects.

Three types of substrate involve special consideration:

1. Fermentation of *pentoses.*
2. Fermentation of the *sugar alcohols* depends on the development of specific dehydrogenases, which are induced. When *E. coli* strain B is grown on sorbitol, it can dehydrogenate 6 phospho-D-sorbitol but not when grown on mannitol or dulcitol (Wolff and Kaplan, 1956). When grown on mannitol (or to a less extent on sorbitol) it can dehydrogenate 1 -phospho-D-mannitol, and this dehydrogenase can be readily prepared from the cells. Thus, there are two different inducible dehydrogenases, and probably a third for dulcitol. Special kinases for phosphorylating the alcohols probably exist too, like the glycerokinase of Aerobacter (Rush *et al.*, 1957) which enables those strains that possess it to form phosphoglycerol; those that do not contain it dehydrogenate the glycerol to dihydroxyacetone like Acetobacter.
3. The fermentation of *uronic acids,* as worked out by Ash well and co-workers (1960), proceeds via the 2 keto-3-deoxygluconic acid route. Glucuronic and galacturonic acids are first isomerised to the keto forms, fructuronic and tagaturonic acids (by an isomerase whose formation is induced by growth on either acid). The keto groups are reduced by DPNH with specific dehydrogenases, and the resulting mannonic and altronic acids then lose H_2O to give 2 keto-3-deoxygluconic acid.

This is phosphorylated and cleaved to pyruvate and phosphoglyceraldehyde. *E. coli, A. aero genes,* and *Serratia marcescens* behave similarly.

Mechanism of Formation

Hydrogen and Formic Acid: The first characteristic in which this complex fermentation differs from those in the three preceding chapters is the production of hydrogen. The key to this is given by the typhoid bacteria, which produce formic acid instead. Having regard to the biological similarities of these bacteria to *E. coli* and the general similarities of the two fermentations in other respects, Harden (1901) and also Pakes and Jollyman, independently, correctly deduced that the hydrogen arises from formic acid. *E. coli* also produces small amounts of formic acid, and indeed Blackwood *et al.* (1956) found that if the *pH* is maintained at 7.8 by adding ammonia, *E. coli* forms much formic acid and very little H_2. Whether this is an effect of *pH.* or of excess ammonia is not clear. AH the analyses show that the hydrogen formed is about equivalent to the CO_2, so that the reaction evidently is:

$$HCOOH \longrightarrow CO_2 + H_2$$

The first clue to the origin of formic acid was Neuberg's finding (1914) that it arose from the fermentation (by mixed cultures) of pyruvic acid:

$$CH_3COCOOH + H_2O \longrightarrow CH_3COOH + HCOOH$$

Oxalacetic acid similarly yields acetic, formic, and CO_2 (although the mechanism of this reaction is far from clear):

$$HOOCCH_2CO\text{-}COOH + H_2O \longrightarrow CH_3COOH + HCOOH + CO_2$$

Pyruvic acid reacts directly with co-enzyme A to acetylate it at the S atom and liberate formic acid:

$$CH_3COCOOH + HS\text{-}(CoA\text{-}enzyme) \longrightarrow CH_3CO\text{-}S\ (CoA\text{-}enzyme) + HCOOH$$

The enzyme is transacetylase.

The acetyl-co-enzyme A rapidly takes up phosphate from the solution to produce acetyl phosphate:

$$CH_3CO\text{-}S(CoA\text{-}enzyme) + HO\text{-}H_2PO_3 \longrightarrow CH_3CO\text{-}O\text{-}H_2PO_3 + HS(CoA) + enzyme$$

Both these reactions are reversible (Stadtinan *et al.*, 1951), but the excess of pyruvate drives them from left to right and the overall reaction is thus:

$$CH_3COCOOH + H_3PO_4 \longrightarrow CH_3COO\text{-}H_2PO_3 + HCOOH$$

Its reversibility was shown in cell-free extracts of *E. coli by* fermenting pyruvate in the presence of acetate or formate containing C^{13} in the carboxyl group (Lipmann, 1946). After a short period $HC^{13}OOH$ gave rise to $CH_3COC^{13}OOH$, the rate depending on the concentration of phosphate present. Cocarboxylase (thiamine pyrophosphate), cysteine, Mn^{++}, and Mg^{++} are needed to catalyse the reaction.

Unlike formate, isotopic acetate is not converted to pyruvate unless ATP is present. The reason for this is that acetate does not directly react with the enzyme. The function of ATP is to form the adenylate of the enzyme, which then exchanges with co-enzyme A to form an enzyme-co-enzyme complex. It is this that reacts with acetate (Jones, Lipmann, Hilz, and Lynen, 1953). The resulting acetyl-enzymec transfers the acetyl group to co-enzyme A to form acetyl-Co A, which is then converted to acetyl phosphate as shown above.

The reactions are thus:

1. $CH_3COCOOH + HS\ CoA \longrightarrow CH_3CO\text{-}S\ CoA + HCOOH$
2. a. $Ad \rightarrow (P)(P) + Enz.OH \longrightarrow Enz. \rightarrow Ad + \rightarrow OH$
   b. $Enz. \rightarrow Ad + HS\ CoA \longrightarrow Enz.S\ CoA + Ad \rightarrow H$
   c. $CH_3COOH + Enz.S\ CoA \longrightarrow CH_3CO\text{-}S\ CoA + Enz.OH$
3. $CH_3CO\text{-}S\ CoA + \rightarrow OH \longrightarrow CH_3CO\text{-}O \rightarrow + HS\ CoA$

1 + 3. $CH_3COCOOH + \rightarrow OH \longrightarrow CH_3CO\text{-}O\rightarrow$ + HCOOH

2 + 3. $CH_3COOH + \rightarrow OH \longrightarrow Ad$ © → → CH_3CO-O→ + Ad→H Sum, (reversing [8]):

$$CH_3COCOOH + \rightarrow \rightarrow OH + Ad \rightarrow H \longrightarrow CH_3COOH + HCOOH$$

where Ad = adenosine and © = phosphate, so that Ad^{rac} = ATP, and ©©OH = pyrophosphate. The reverse of equation (8) yields free acetic acid.

In sum 1 therefore, this fermentation rests on the conversion of pyruvic acid to acetic and formic acids, with the intervention of co-enzyme A, ATP, and phosphate, and the decomposition of formic acid to CO_2., and H_g.

Alcohol: Although the splitting of pyruvate explains the production of formic acid, it does not by any means explain the whole fermentation, because alcohol is formed roughly equivalent to the acetic and formic acids. This alcohol cannot be formed as in yeast by decarboxylation of pyruvate and reduction of the acetaldehyde, since if so an equal quantity of CO_2 would be formed. We have seen that CO_2 equivalent to the H_2 comes from formic acid, and as a rule the ratio CO_2: H_2 for *E. coli* is close to 1. (With Aerobacter it is nearer 3, but there are special reasons). Besides, pyruvic dccarboxylase is not present, because. The value of 1 for the Ct>2: H_2 ratio has, however, not always been obtained. For example, the data of Grey and Young (1921), obtained under anaerobic conditions, lead to the following quantities, per 100 moles of glucose fermented:

CO_2	119
H_2	34.2
HCOOH	37
CH_3COOH	133
C_2H_5OH	46
Lactic acid	8.2
Succinic acid	0.5

Compared to those quoted above, they show much more CO_2 and acetic acid and much less lactic and succinic acids. Variations in yields of the acids may be ascribed to the *pH*, but the high ratio of CO_2 to H_2 conflicts not only with Scheffer's data but also with common experience of many workers, cell-free extracts do not act on pyruvate (Dawes and Foster, 1956). On the other hand acetaldehyde is formed in the fermentation, as shown both by adding bisulphite (Neuberg and Nord, 1919) and by sampling and testing during the first hour (Dawes and Foster, 1956). There is also a reversible DPN-linked alcohol dehydrogenase present (Still, 1940), which could convert acetaldehyde to alcohol. These conflicting considerations were harmonised by the following observations (Dawes and Foster, 1956):

1. Fluoride, cyanide, and iodoacetate, all of which inhibited pyruvate formation, also stopped the production of ethanol; arsenite, which caused an accumulation of pyruvate by preventing its metabolism, also stopped the production of ethanol. Thus, ethanol is probably formed from pyruvate.
2. A mutant that requires pantothenate for growth, and is therefore presumably deficient in co-enzyme A, metabolises pyruvate and produces ethanol, in proportion to the amount of pantothenate added. Thus ethanol formation requires co-enzyme A.
3. The cell-free acetaldehyde dehydrogenase of *E. coli*, using DPN, acts only in presence of co-enzyme A. With DPNH it reduces acetyl-co-enzyme A; i.e., it establishes the equilibrium:

$$CH_3CHO + DPN+ + HS\text{-}CoA \longrightarrow CH_8CO\text{-}SCoA + DPNH + H^+$$

It follows that the ethanol formed by *E. coli* comes from acetyl-CoA via acetaldehyde and subsequent reduction.

The yield of alcohol in the natural fermentation is more constant than that of the acids. The wide variation in yields of the latter may well be due to differences in pH, which often has not been controlled. With increasing alkalinity the yield of acetic and formic acids goes up and that of lactic acid goes down. The Ho gas has been considered as coming from formic acid and has therefore been added to the formic acid found as such. The production of lactic acid or of formic plus acetic acid represent, roughly speaking, alternative fates for pyruvic acid, while the production of alcohol remains constant. In general, the effect of *pU* resembles that noted with the heterofermentative lactic organisms; alkalinity lowers the yield of lactate and increases the yield of acetate.

Lactic Acid: Cell-free enzyme preparations from Escherichia and Aerobacter can split hexosediphosphate into the two triose derivatives, phosphoglyceraldehyde and phosphodihydroxyacetone, and can also establish equilibrium between these two products (Utter and Werkman, 1941-1942). The dehydrogenation of phosphoglyceraldehyde to phosphoglyceric acid (Still, 1940) and the conversion of phosphoglyceric to phosphopyruvic acid, as well as the transfer of phosphate from this product to adenylic acid (Utter and Werkman) have been similarly established. The bacteria thus produce pyruvic acid in the "standard" manner. When it is considered that they can also dehydrogenate lactic acid, and that such dehydrogenations are reversible, it is evident that the lactic acid produced in this fermentation arises in exactly the same way as in the lactic fermentation. The difference between the *Enterobactenaceae* and the homotermentative Lactobacilli lies in the fact that the former have an additional method of dealing with pyruvic acid (namely, its conversion to formic and acetic acids).

Succinic Acid and the Fixation of CO_2 as a Factor in Growth

The fact that succinic acid (containing 4 carbon atoms) can be produced by *E. coli* from glycerol (containing 3 carbon atoms) just about as freely as from glucose (Grey, 1924) indicates that it arises by the addition of CO_2 to pyruvate and subsequent reduction of the oxaloacetate formed. This was proved by using cell- free extracts of *E. coli,* with isotopic CO_2 (Kalnitsky and Werkman, 1943-1944). These preparations are free of cocarboxylase and hence do not destroy pyruvate; they thus convert added CO_2 to succinate. ATP and biotin are essential for the reaction, as well as either Mn or Mg. The succinate produced in the formic fermentation, therefore, comprises part of the total yield of CO_2. Werkman and his co-workers have elucidated several other reactions by which CO_2 is "fixed" into organic acids. These may not account for an appreciable consumption of CO_2 in the fermentation, but they indicate how widely CO_2 enters into combination:

1. The first step in the oxidative decarboxylation of pyruvate to acetate and CO_2 is reversible, so that acetyl-CoA can absorb CO_2, to form pyruvate and thus eventually citrate:

   TPP

   $CH_3CO\text{-}C^{13}OO\text{-} + HS\text{-}CoA + DPN+ \longrightarrow CH_3CO\ S\text{-}CoA + C^{13}O_2 + DPNH$

   This was first shown by incubating washed cells or a cell-free preparation of *M. pyogenes* var. *aureus* with pyruvate and $C^{13}O_2$, when the pyruvate came to contain C^{13} in its carboxyl (Wiken *et al.*, 1947; Watt and Werkman, 1950). A similar fixation occurs with a cell-free preparation from *Cl. butylicum* which evolves H_2 (Wilson *et al.*, 1948):

   TPP

   $CH_3CO\text{-}C^{13}OO^- + HS\text{-}CoA \longrightarrow CH_3CO\ S\text{-}CoA + C^{13}O_2 + H_2$

   In both cases ATP promoted the reaction as would be expected; co-enzyme A was not added but would have been present in the preparation along with cocarboxylase.

2. The oxidative decarboxylation of a-ketoglutarate to succinate and CO_2 is also to some extent reversible. A cell-free enzyme from *E. coli* will decarboxylate a-ketoglutarate, and the resulting succinate will

accumulate if maloriale is added to inhibit its oxidation. If now $C^{13}O_2$ is introduced and the reaction stopped, the a-ketoglutarate contains C^{13} in the carboxyl adjacent to the C~0 group. The fixation is promoted by ATP and involves succinyl phosphate as intermediate.

3. CO_2 is fixed in the formation of purines and pyrimidines. With *A. aerogenes*, isotope analysis shows that the 4, 5, and 7 positions of its adenine and guanine come from glycine, the 2 and 8 positions from formate, and the 6 position from CO_2 (Sutton and Werkman, 1953). The N at position 1, 3, and 9 comes from glutamine.
4. The formation of carbamyl phosphate, and therefore the synthesis of citrulline, arginine, and pyrimidines involves CO_2 fixation.
5. Serine is formed from glycine, either with CO_2 or with formate (Lascelles *et al.*, 1951). Folic acid and pyridoxal are needed as co- enzymes. All of these reactions using CO_2 account for the fact that many bacteria require CO_2 for growth. If cultures are very vigorously aerated with CO_2-free air, growth may be reduced; yeast extract will then substitute for CO_2 (Lwoff and Moriod, 1947). Anaerobic growth is often dependent on CO_2 also, especially among the coliforms. In the absence of CO_2, growth of Escherichia and Aerobacter is reinstated by many organic acids, but for *Serratia marcescens,* the replacement of CO_2 requires asparagine, arginine, uracil, adenine, and guanine; each alone gives rise to a little growth and together they replace 93 per cent of the fixed CO_2 (McLean and Purdie, 1952). Evidently, therefore, CO, is fixed in dicarboxylic acids, purines, and also in arginine. Thus CO_2 fixation provides an important part of nutrition.

Assimilate

The bacteria grow during the fermentation. In oxidations, approximately a third of the carbohydrate decomposed appears as *E. coli* cell substance (Cook and Stephenson, 1928). In fermentation the assimilation is much less, although it is not negligible. A complete accounting of all carbon in the form of fermentation products is therefore not to be expected. The assimilate has approximately the composition of carbohydrate. In the scheme below we have shown per cent mole of pyruvic acid from a total of 4 moles, i.e., 12.5 per cent, diverted to cell substance.

C. Overall Scheme

The above arguments give rise to a general scheme for the formic fermentation of glucose roughly. The scheme accounts fairly well for the quantitative distribution of products commonly found. Intermediate steps, such as the formation and degradation of acetyl-CoA, have been omitted. The relative amounts of the various products will necessarily depend on the relative activities of the enzymes taking part and thus will vary somewhat with the particular organism and the growth conditions.

Aerobacter Modification

Among the bacteria that can ferment lactose to formate, there are those like *E. coli* that produce CO_2 and H_2 in a ratio about 1, and others, including Aerobacter and Serratia, in which the ratio is 3 or much higher, approaching 100 with some strains of Serratia and also with *Bac. sublilis.* The presence of two such clearly defined groups has long been recognised (Harden, 1905; Harden and Walpole, 1906) and is of great importance because organisms of the second type are soil forms, while *E. coli* is intestinal in origin. Establishment of the presence of a formicfermenting organism in a sample of food or water is thus only of significance for public health (i.e., showing faecal contamination) if it is clear that the organism is not a native soil form.

Formation of Acetoin

The principal difference lies in the formation by Aerobacter of a new substance, acetoin or acetvlmethvlcarbinol.

CH,COCHOHCHa (Harden, 1906; Harden and Walpole, 1906). Pyruvate combines with thiamine pyrophosphate at the 2 position of the thiazole ring, but on losing its CO_2 the acetyl residue, instead of being passed to lipoic acid as described there, is passed to another molecule of pyruvate to form D-a-aceto-lactate; this is decarboxylated:

$$
\begin{array}{l}
COCH_3 \\
\mid \\
[CH_2CO\text{-}H] + CH_2CO\text{-}COOH \rightarrow CH_8C_2COOH \rightarrow CH_3CHOHCOCH_3 + CO_2 \\
\mid \\
OH
\end{array}
$$

Thus, each molecule of acetoin formed releases 2 CO_2. The enzyme appears to be the same as pyruvic decarboxylase and to have *two* sites, for holding on to two acetyl residues (Juni, 1961). Its *pH* optimum is about 6, which agrees with the optimum *pH.* of acetoin formation. The acetolactate may have an alternative fate, because there is a second enzyme system, of optimum *pH* 7.8, producing it, and this one is inhibited by valine (Halpern and Umbarger, 1959). This system is thought to be the one by which valine is *formed,* by way of a molecular rearrangement and transamination. Acetoin can be oxidised to diacetyl, and it can also be reduced to 2,3-butanediol, and it is this compound which is mainly found in the fermented liquor.

It will be remembered that every mole of pyruvate formed involves the simultaneous reduction of DPN, and the 2(H) is doubtless at once transferred to acetoin, just as it would be to acetaldehyde in fermentation by yeast. If an Aerobacter fermentation liquor is made strongly alkaline and exposed to air, it slowly turns bright pink (Voges-Proskauer reaction). Addition of *a*- naphthol and creatine hastens the reaction and makes it more sensitive, since it depends on condensation of the diacetyl with traces of creatine or guanidine derivatives in the peptone. The Voges-Proskauer reaction is one of the four standard tests used for differentiating the Aerobacter and coli fermentations. The four tests, abbreviated, are: Indole (formed by *E. coli* from tryptophan), Methyl red (turned red by *E. coli,* due to the acetic acid formed), Voges-Proskauer reaction Citrate (utilised as sole carbon source by Aerobacter) Because of the acetic acid produced, the *E. coli* fermentation typically reaches pH 4 in about 24 hours, while that with Aerobacter goes only to pH 5.3 so that methyl red remains orange-yellow in it.

Like the other fermentation types, that by Aerobacter is strongly pH sensitive. The formation of acetoin and 2,3-butylene glycol takes place only on the acid side of pH 6.3 (Watt and Werkman, 1951). At more alkaline *pH* the products are mainly lactate, acetate, and ethanol, so that the fermentation more nearly resembles that by *E. coli.* Diacetyl has another importance. It possesses a characteristic odour, like that of good butter. It is indeed probably responsible for the aroma of butter (van Niel *et al.,* 1929) and is sometimes added in minute amounts to margarine to give it a buttery flavour. Its production in soured cream is largely due to *Sc. cremoris* and *Sc. diacetilactis,* which form it from citric acid.

General Scheme

Because so much pyruvic acid is converted to acetoin (at acid pH) there is less of it to be converted into acetic acid (cf. the methyl red test). Furthermore, since each mole of acetoin corresponds to the formation of 2 moles of CO_2, the total CO, derives partly from formic acid and partly from this extra source. Hence the ratio CO_2:H_2 is much above 1. Typical analyses of the products from 100 moles of glucose fermented by *"Aerobacter indologenes"* (a variant of *A. cloacae).*

Allowing 2 CO_2 for each mole of 2,3-butanediol (i.e., 133 CO_2), 39 moles of CO_2 remain; this just corresponds to the 36 moles of H_2. Hence the remaining CO_2, like the H_2, came from formic acid. In general, in these fermentations, the CO_2 plus the succinic acid (which represents additional CO_2 that has been fixed) should be equal to the H_2 plus twice the butanediol. In the data for Serratia (from Neish, 1950), very little H_2 is evolved, and the yield of CO_2 is thus just twice that of the butanediol. Unlike Aerobacter, Serratia's optimum *pH*. for butanediol formation is about 7, and presence of $CaCO_3$ allows good yields (Bahadur and Dube, 1958). The data for *Bac suhtilis* agree well. Only the Escherichia forms acetic acid, although the other organisms shown may form a little in the earlier stages. The Aerobacter data were obtained after 209 hours; at 31 hours there were 7 moles of acetic acid formed, but even then there were 21 moles of formic acid.

It is evident, therefore, that ethanol is appearing instead of acetic acid, in other words, that the acetyl group is being reduced before it can be hydrolysed to acetic acid. The appearance of small amounts of acetate early in the fermentation and its subsequent disappearance indicate that acetate is being condensed with formate to pyruvate. It is important to observe that in this condensation and its sequel the carboxyl of acetic acid is reduced. In other words, *carboxyl* carbon of acetate is converted to *carbinol* carbon of butylene glycol.

The general course of the fermentation and the differences from that of Escherichia are all seen to be results of the formation of acetoin. It is of interest that the methyl red test, the CO_2:H_2 ratio, and the Voges-Proskauer reaction were all used by public health bacteriologists to distinguish the two types of organism long before the underlying mechanism was understood.

It should be added that *E. coli* also can produce small amounts of acetoin, especially in vigorously aerated cultures containing 2 per cent $NaHCO_3$ (Reynolds and Werkman, 1937). If the reaction products are distilled and the test carried out on the distillate, the sensitivity is higher, and even at the ordinary *pH a* positive reaction is given (Kluyver and Molt, 1939); with some strains the yield approaches 1 per cent (Blackwood *et al.*, 1956). However, *E. coli* does not normally show the test on the fermentation liquor directly.

Fermentation of Pentoses

When pentoses are the substrate, the products are often the same as with hexoses. In this respect the gram-negative rods differ from the lactic bacteria, in which pentoses give quite a different type of fermentation from hexoses. In the lactic organisms, as also in *Ps. lindneri*, the pentose is split into a 3-C and a 2-C fragment, the one yielding lactic acid and the other alcohol or acetic acid. Since the Enterobacteriaceae convert pentoses to the *same* products as hexoses, the 2-carbon fragment must be somehow resynthesised to triose. This is in fact what occurs. The resynthesis is brought about by the enzymes, transketolase and transaldolase, and the co-enzyme, thiaminepyrophosphate. Aerobacter provides the best bacterial example of this remarkable series of reactions, which were worked out mostly on peas, spinach, and liver extracts.

In the above scheme, if the 1-carbon of the ribose (C) is an isotope, three of the resulting trioses will be unlabelled, one will be labelled in the 1 position, and one labelled in both 1 and 3 positions. As a result one of the 5 molecules of pyruvate will have isotope in the methyl group, and one will have isotope in both methyl and carboxyl groups. Exactly this pattern was found when *A. aerogenes* fermented D- and L-arabinose labelled in the 1 position (Neish and Simpson, 1954); the C^{14} activity in the CH_4 groups of the lactic and acetic acids, butanediol, and ethanol was 40 per cent of that in the original C-1 of the sugar, while in the COOH of the lactic acid, in the formic acid, and in the CO_2 the activity was just one-half as great (i.e., 17 to 23 per cent). Similar though less complete evidence indicates that *A. cloacae* and *E. coli* use the same route (Simpson and Wood, 1956; Nutting and Carson, 1952).

The methylpentose rhamnose may be metabolised in the same way, since the first step is conversion to the keto-form rhamnulose, which is then phosphorylated (Englesberg and Baron, 1958). When a hexose molecule is fermented, the net gain in ATP is 2 molecules, or per cent ATP per carbon atom. When pentose is metabolised by this route, 3 molecules (requiring 3 ATP) give 5 trioses of which two need phosphorylation (= 2 ATP), and each of the 5 then yields 2 ATP; in all, 5 ATP result from 3 pentoses or per cent ATP per carbon atom. It would be expected, therefore, that growth on pentose would be just as rapid as on hexose, and in general this expectation is justified. *A. aerogenes* grows well on six different aldopentoses.

Acetoin Formation by Other Bacteria

Aerobacillus polymyxa, as mentioned above, converts added acetaldehyde to acetoin and butylene glycol (Stahly and Werkman, 1942). This organism forms alcohol and acetone, as well as acetoin, from glucose and other sugars, from sugar alcohols, and from pectins. Its close relative *Aerobac. macerans* does not form acetoin. *Lactobacillus plantarum*, one of the streptobacteria, produces acetoin and CO_2 from pyruvate (Rowatt, 1951). The luminous pseudomonads, *Photobaderium*, carry out a fairly typical formic fermentation and produce varying amounts of acetoin (Doudoroff, 1942).

Among other organisms forming traces of acetoin are certain yeasts, especially *Pichia indica*, in which the acetoin may reach 2.5 per cent of the sugar fermented (Bahadur and Ranganayaki, 1958); here the production is promoted by adding zinc. More remarkable are the high yields of acetoin and butanediol produced by *Bac. subtilis*. Although this sporeformer is strongly aerobic, it can ferment glucose to the following products, in order of decreasing concentration; CO_2, butanediol (and acetoin), lactic acid, glycerol, ethanol, and a little formic, succinic and acetic acids (Neish, 1953). Except for the reduction of about one-sixth of the triose to glycerol, the fermentation resembles that of Serratia above. When carried out in air, the yield of butanediol can be as high as 87 moles per 100 moles glucose fermented (Blackwood and Simpson, 1950). Indeed it was observed as long ago as 1906 by Harden that *Bac. subtilis* and *Bac. mesentericus* (a closely related form) gave strong Voges-Proskauer reactions. Formation of acetoin by the heterofermentative lactic bacteria, *Leuconostoc citrovorum*, was mentioned above. Here carbon atoms 1, 2, and 3 yield alcohol and CO_2, while 4, 5, and 6 yield pyruvate and thence lactate. This pyruvate together with pyruvate derived from citric acid in the milk is converted to acetolactate and thence acetoin. Added citrate or pyruvate yields acetoin more readily than glucose does (Busse and Kandler, 1961). Several Pseudomonads, especially *Ps. fermentans* and *Ps. trijolii* (=5. *herbicola)*, produce acetoin from sugars; the former attacks lactose and thus, except for its polar flagellation, is closely related to Aerobacter. The latter shows its relationship to Aerobacter in another way, i.e., by the ability to form much gum on sugar media.

Enzymes Concerned with Hydrogen Gas

Although hydrogen comes almost always from formate, formate can be decomposed in several ways:

1. Washed bacteria incubated with formate and methylene blue reduce the dye (Quastel and Whetham, 1925):

   $$HCOOH + MB \longrightarrow MBH_2 + CO_2$$

   The enzyme, which is present in Escherichia, Aerobacter, and others of the group, is called *formic dehydrogenase.* Instead of methylene blue, it can reduce nitrate (in the presence of the enzyme nitratase) and it can also react with cytochrome *c.* This last property is explained by the fact that formic dehydrogenase, like succinic and lactic dehydrogenases, is closely bound to a cytochrome of the *b* type. Recently such a complex has been extracted from *E. coli;* it contains both cytochrome b_l and a flavin, as well as 30 per cent RNA (Wrigley arid Linnane, 1961). In the spectophotometer formate could be seen to reduce *both* the cytochrome and the flavin. Succinic dehydrogenase arid nitratase activity were absent. In another preparation (Itagaki *et al.*, 1961) reduction of the cytochrome requires participation of a lipid fraction, replaceable by vitamin K_a; without this, the formate will reduce methylene blue but not the cytochrome.

2. Washed *E. coli* can also reduce methylene blue with hydrogen gas (Stephonson and Stickland. 1932), which means that hydrogen is being brought into chemical combination, as by palladium black or Raney nickel

$$H_2 \longrightarrow 2\ H^+ + 2\ e^-$$

This enzyme, which is called *hydrogenase* is apparently a hemin derivative; it is inactivated reversibly by oxygen and by NO (Fisher *et al.*, 1954); it is also inhibited by CO though not by cyanide (Joklik, 1950). Its activity depends on a SH group and is inhibited by mercury compounds. In *Ps. saccharophila* it is present in two forms, one of which reduces DPN and the other, methylene blue. In the cell- free preparation these could be separated on the centrifuge (Bone, 1960). The enzyme occurs widely, especially in Azotobacter, in purple bacteria, and in those "Knallgasbakterien" that can grow by oxidising hydrogen, In the purple *Rh-ps. palustris* it is in particles, probably chromatophores; like the much stronger *E. coli* enzyme, it is inhibited by CO but not by HCN (Izawa, 1962).

3. The evolution of H_2 by *E. coli* does not quite follow either of these patterns. It proceeds optimally at *pH*, 6 and is inhibited by cyanide, 10 M KCN giving 50 per cent inhibition (Stephenson and Stickland, 1932). For these reasons Woods (1936) at first ascribed it to a separate enzyme, hydrogenlyase, catalysing the reversible reaction: $HCOOH \rightarrow H_2 + CO_2$ Evidently this enzyme, combined with formic dehydrogenase, would allow Ho gas to reduce methylene blue, in other words, would duplicate the action of hydrogenase. Hydrogenlyase is developed in *E. coli* as a function of its growth medium. Grown on plain broth, the organisms produce no hydrogen when suspended in glucose. They do, however, contain hydrogenase and formic dehydrogenase, i.e., they can reduce methylene

blue with H_2 or with formate. If the bacteria are grown briefly on broth containing 1 per cent formate they then contain the hydrogenlyase. Non-gasproducing strains of *S. paratyphosa* can also develop the enzyme after growth in formate medium (Pot and Tasman, 1932-1933). The other two enzymes are not induced under these conditions. If the formate contains deuterium instead of hydrogen, the rate of gas evolution is halved, but the rate of methylene blue reduction is not changed (Abei and Frei, 1957).

The special nature of the "lyase" was brought out unexpectedly by experiments with iron-deficient cells (Waring and Werkman, 1944).

E. coli grown on broth very low in Fe contain little or no catalase or peroxidase and show no visible cytochrome bands; they also have no activity of formic dehydrogenase, hydrogenase, or hydrogenlyase. On *partially* Fe-deficient media the cells have formic dehydrogenase and hydrogenase, but in fermenting glucose they produce no H_2, producing instead large amounts of formic acid; hence they have no hydrogenlyase. Since there is still enough Fe present for growth, it seems that the "lyase" is "more easily suppressed by Fe deficiency than any other system studied." Since also the "lyase" is powerfully inhibited by KCN, Waring and Werkman (1944) postulate that liberation of H_2 from formate requires (1) formic dehydrogenase, (2) hydrogenase, and (3) an intermediate electron mediator, containing Fe, probably similar to cytochrome. This makes unnecessary a separate "lyase" enzyme but accounts for the special nature of the system.

Glycerol and the E. Freundii Modification

Glycerol is readily fermented by the *Enterobacteriaceae.* With *E. coli* the products are mainly alcohol, lactic acid, formic acid, and extra CO_2 (Harden, 1901); with Aerobacter they are mainly alcohol and acetic and formic acids (Braak, 1928). However, when carried out under anaerobic conditions, the fermentation comes to a stop long before all the glycerol is used; addition of more peptone or yeast extract will start it again, but only for a time. Braak (1928) deduced that the dehydrogenation of glycerol $CH_2OHCHOHCH_2OH \longrightarrow CH_2OHCHOHCHO +$ 2(H) was the limiting reaction arid that the peptone was needed to act as hydrogen acceptor. Acetaldehyde acts similarly as an acceptor (increasing the yield of ethanol), or the fermentation could be carried out in air, oxygen being then the acceptor. (The subsequent stages from glyceraldehyde, or more probably phosphoglyceraldehyde, are doubtless essentially as in the fermentation of sugar.) It was, however, observed by Freund in the 1880's that some fermentations of glycerol gave rise to trimethylene glycol, $CH_2OH\text{-}CH_2\text{-}CH_2OH$, and Voisenet (1918) isolated organisms carrying out this process, although he was mainly interested in the fact that they produced acrolein, which accounted for a bitter taste in infected wine. Braak, in his enrichment cultures, obtained two organisms that fermented glycerol rapidly with no added hydrogen acceptor. Evidently, therefore, the glycerol itself was being used as hydrogen acceptor, and trimethylene glycol was identified in the residues:

$$CH_2OHCHOHCH_2OH + 2(H) \longrightarrow CH_2OHCH_2CH_2OH$$

It will be seen that the second strain produces more formic acid and alcohol, but less trimethylene glycol, than the first. Both were termed *B. freundii* (now *E. freundii*). Some organisms reducing glycerol to trimethylene glycol are found among the lactobacilli (Sobolev and Smiley, 1960) and *Acrobactcr aerogenes* (Mickelson and Werkman, 1940). The *Lactobacillus* species can obtain the 2(H) for reducing the glycerol either from glucose, or from the formation of 3-hydroxypropionate. Small amounts of acrolein are produced as well, non-enzymically, so that in all probability the intermediate for all three products is 3-hydroxypropionaldehyde.

E. freundii, on the other hand, derives the 2(H) direct from glycerol to form trimethylene glycol, and in strain 19 of the table we see that each mole of trimethylene glycol equals the sum of lactic, succinic, and acetic acids and alcohol. Strain 20, however, ferments additional glycerol to alcohol and formic acid.

Dihydroxyacetone is fermented in a comparable way. *E. coli* attacks it only very slowly but does so rapidly after being grown in its presence (Virtanen *et al.*, 1930), most probably due to induced formation of the kinase which phosphorylates it. The fermentation then yields mainly one mole of glycerol and one mole each of acetic and formic acids. Evidently 1 mole is reduced, while the other is phosphorylated, converted to phosphoglyceric acid and thence to acetic and formic acids.

Butyric and Butylic Bacteria

The fermentations considered have all been the work of facultative anaerobes. The butyl-butyric-acetone fermentations, on the other hand, are brought about by Clostridia, which are strict anaerobes, and in this chapter it will be convenient to consider also some other processes due to clostridia.

The fermentations are characterised by production, from sugars or related compounds, of butyric acid together with one or more of the following: butyl alcohol; the C_3-compounds acetone and isopropyl alcohol; acetic acid; formic acid; ethyl alcohol; usually also neurogen and some CO_2. Butyric acid is produced not only by clostridia but also by some anaerobic Neisserias, by *Zymosarcina maxima*, and (though not from sugar) by the non-sporeforming anaerobic rods called *Butyribacterium*. Butyric acid is formed also from threonine by a special fermentation due to *Clostridium propionicum*.

Butanol is not produced by other microorganisms, but isobutanol is produced by yeast as a constituent of fusel oil. Acetone can be produced by certain facultative anaerobes, and the other products are of course common to many fermentations. The butyric fermentation was discovered in 1861 by Pasteur, who named the causative organism the *vibrion butyrique*, saw that it was killed by air, and coined the term *anaerobic* for it. Fitz in the 1880's and both Gruber and Grimbert later studied the fermentations but did not distinguish between the butyric acid and the butanol- producing forms. Pure cultures of a butyric acid-producing form were obtained from soil in 1895 by Winogradsky, the main purpose of whose work was the study of nitrogen fixation.

Winogradsky's organism, named *Cl. pastorianum*, produced butyric acid and is readily obtained by fermenting potato slices with a little $CaCO_3$. He noted, however, the formation of small amounts of butyl and ethyl alcohols. Bacteria producing appreciable amounts of butanol were first obtained in pure culture by Beijerinck (1894), who clearly recognised the difference between them and the butyric organisms; his name *Granulobacter butylicum* was based on presence of a starchlike substance in the cell, staining blue with iodine, which he called *granulose*. From their ability to attack starch vigorously, forming a clear zone on starch agar plates, the organisms have also been called *Amylobacter*.

Another of Beijerinck's isolates attacked pectin as well as starch and was called *Granulobacter pectinovorum*. The name *Granulobacter* was later dropped in favour of *Clostridium*, an old term referring to the swollen rods containing spores. An organism with the useful ability of making both butanol and acetone was isolated by Fernbach in 1911 and another by Weizmann in 1915.

Because the latter, *CL. acetobutylicum*, also known as "BY" bacterium, was the more actively fermenting strain and would attack corn, it has become the basis of the butanol-acetone industry. The setting up of this industry in the United States resulted from the establishment of a plant in Canada during World War I, when acetone was in great demand as a solvent for cordite. At present the butanol and butyl esters formed from this fermentation are just as much in demand as the acetone, but synthetic chemicals provide serious competition. Aerobic bacilli making acetone together with ethanol and acetic acid were isolated by Schardinger (1905) under the name of *Bac. macerans*.

A related form, which produces 2,3-butaneodiol, was first studied by Prazmowski in 1880 and by Beijerinck (1894) and named *Clostridium* (or *Granulobacter) polymyxa* because morphologically like the butanol bacteria, and making *much gum* (= *polymyxa).* Neither of these produces more than a trace of butanol. Because these organisms produce a fermentation similar to that of Aerobacter they were made into a separate genus Aerobacillus by Donker (1926).

A group of non-sporeforming gram-positive rods which form acetic and butyric acids and CO_2, but no hydrogen, was characterised by Barker and Haas (1944) as a new genus, *Buiyribacterium.* These organisms are probably more nearly related to the heterofermentative lactobacilli than to the clostridia.

The clostridia as a group are morphologically quite alike. They are mostly gram-positive rods; in the early stages of a fermentation when traces of oxygen are still present, they are medium-sized and actively Both Fernbach and Weizmann patented their organisms, as both were working in industry (Strange and Graham, Ltd., England), though Weizmann had left the firm when he isolated his organism. Dr. Weizmann afterwards became interested in Zionism, of which he was World President for many years, and the patent profits on *Cl. acetobutylicum* were largely used to promote the Zionist cause.

As President of Israel from 1949 to 1952, Dr. Weizmann was one of the few microbiologists to become successful in politics motile; in later stages, when conditions become really anaerobic, they become larger and either boat-shaped or drumstick-shaped, with good-sized spores (1 X 2 *p.)* inside. They differ from one another in size, spore position (central or terminal), heat resistance of the spores, and in their ability to fix nitrogen. On physiological grounds it is convenient to regard the butyric Organisms as falling into five groups, though the border lines are not too sharply drawn.

1. Butyric bacteria *(Cl. saccharobutyricum, pasteurianum,*

and others) make butyric and acetic acids from glucose; most of these also produce traces of butanol or ethanol, depending on the growth conditions.

2. Butylic bacteria *(Cl. butylicum, acetobutylicum,* and others) convert sugars to butanol in good yield, along with either acetone or isopropanol.
3. Caproic acid bacteria *(Cl. kluyveri)* do not attack sugars, but convert certain acids and ethanol to caproic acid, $C_7H_{15}COOH$, and other fatty acids, with butyric acid as a probable intermediate. Other organisms, in groups 1, 2, and 5, also form some caproic acid.
4. Aerobacilli *(Bac. macerans, poly myxa, acetoethylicum)* make acetone, 2,3-butylene glycol, and ethanol, but not butanol, and only traces of butyric acid.
5. Certain non-sporeformers make butyric acid with acetic acid *(Butynbactenum)* or with acetic and lactic acid.

Some of the clostridia are not primarily butyric fermenters. These "Peptoclostridia" (including *Cl. sporogenes, tetani, histolyticum,* and *botulinum)* attack amino acids in a characteristic way, in which one amino acid is oxidised and another reduced, with deamination.

Generally they are strongly proteolytic, attacking intact animal tissues, especially muscle. Two of them are remarkable for deaminating y-aminobutyric and 8-aminovaleric acids, hydrogenating one molecule to butyric or valeric acid, respectively, and dehydrogenating a second molecule to NH_2 and acetic acid (Hardman and Stadtman, 1960).

Cl. acidiurici and *Cl. cylindrosporum* are notable for fermenting purines. Some members of this group can ferment sugars to 531 butyric acid, etc., but others do this weakly, if at all. Their main importance is as animal pathogens, because they cause tetanus, gangrene of wounds (especially *Cl. welchii),* and botulism.

Most of the industrial interest has centred on the butylic group (group 2). These are rather homogeneous. Under comparable conditions the fermentation products of the four principal members are quantitatively similar. Cl. elsineum is distinguished by its orange pigment and its vigorous attack on pectins, which makes it important in the retting of flax. Cl. beijerinckii may have industrial application in the fermentation of waste sulphite liquors from paper mills; a little corn meal, yeast extract, and salts added to the liquor make it an excellent medium, and the pentosans in it can be fermented to useful products (Gfeller, 1954).

Table: Neutral Products of the Butanol Fermentations from Glucose

(Medium: Yeast-Water, 2 per cent Glucose; Duration 4 to 5 Days at 30° C or 37°C.)

Organism	*Per Cent of the Fermented Glucose*			
	Butanol	*Acetone*	*Isopropanol*	*Et Hanoi* H_2
Cl acetobutylicum	23.0	7.2	2.4	1.5
Cl beijerincMi	26.7	5.7	2.4	1.3
Cl butylicum	27.6	9.2	2.5	1.0
Cl felsineum	21.0	6.2	6.8	1.5

Cl butylicum makes isopropanol instead of acetone, as shown. Among the butyric group, *Cl. pasteurianum* is distinguished by fermenting mannitol, sorbitol, and inulin, and *Cl. amylobacter* by fermenting pectin. *Cl. saccharobutyricum* does not attack these, but all three can ferment starch, dextrin, and glycogen (Brown, 1937). *Cl. lactoacetophilum* ferments lactate or glycerol only in presence of acetate (Bhat and Barker, 1947-1948). The products of a typical fermentation by a Clostridium of group 1 are compared with those formed by *Zymosarcina maxima,* of group 5. Except for the lactic acid in the latter and correspondingly less acetic acid and CO_2, they are remarkably similar. Some others of group 1, e.g., *Cl. uracilicum* (Campbell, 1957), produce less CO_2 and more acids.

Acetone-butanol Fermentation and its Course

The Influence of pH: The formation of the "neutral solvents," acetone and butanol, takes placeonly in acid media. When the fermentations are allowed to go normally, using sugars or corn mash, the *pH* goes down to 4.5; if chalk is added to hold the *pH* to 7 or 8, the yield of butyric acid goes up roughly four times and that of butanol falls to about a tenth; similarly the acetic acid is doubled and the ethanol halved. The acetone almost disappears (van der Lek, 1930; Bernhauer *et al.*, 1936). The data of van der Lek, using *Cl. butylicum.*

Moles Per 100 Moles Hexose

That butylic organisms behave exactly like those of the butyric group if the acids formed are neutralised. We may conclude that *the butyric fermentation is the fundamental type* and that the butylic process comprises additional enzymes which come into action only when sufficient free acids accumulate. The decreases in the yields of reduced compounds (butanol and ethanol) are accompanied by an increase in hydrogen gas.

Influence of Other Factors

In fermentations on synthetic media, the addition of extracts of corn or other natural products increases the yield of butanol. The alcohol-soluble proteins from several other grains have the same effect. Even on corn mash the addition of potato extract or aqueous soybean extract (Hoshino, 1953) causes a further increase. Much of this effect was traced to aspartic acid, asparagine, or glutamic acid (Tatum *et al.*, 1934-1935). With *Cl. butylicum* on corn mash, asparagine raised the yield of butanol to 17 per cent; however, Yeast extract can be more effective, raising it to 27 per cent. Since the optimum concentration of asparagine is 0.01 M (Brown *et al.*, 1938), it can hardly be itself a precursor of butanol. Metallic ions, especially iron, strongly influence the fermentation.

With *Cl. perjringens,* whose products are mainly butyric and acetic acids with ethanol and H., the absence of iron in the medium changed the products over to lactic acid and only small amounts of other compounds (Pappenheimer and Shaskan, 1944). Carbon monoxide, which combines with iron enzymes, also changes the fermentation to a lactic type, both with butyric organisms (Kubowitz, 1934) and typically butylic ones (Simon, 1947). Since low iron favours riboflavin synthesis, some relation between this and the lactic fermentation may be guessed at. Added iron, on the other hand, depresses the yields, especially of butanol (Imai, 1956).

It seems that molasses heated with Fe-salts forms a fermentation-inhibiting factor, and that this may act by promoting the growth of bacteriophages (Hoshino, 1953), but the full explanation is uncertain. If corn is extracted with dilute acid or alkali, its fermentation is slowed down and cannot be accelerated by adding asparagine or proteins.

In this case the effect has been traced to potassium ions (Davies, 1942), KC1 restoring the acetone yield of *Cl. acetobutylicum* almost to normal. Finally, autoclaving glucose with the yeast extract medium gives much less butanol and acetone than autoclaving it separately (van der Lek, 1930). These and many other causes of variation have been encountered in large-scale applications and in study of the fermentation. Their complete elucidation will require considerable research.

The "Break"

In butylic fermentations the acidity (due to acetic and butyric acids') increases at first and then, aftersome hours, breaks sharply At the same time, much CO_2 is evolved. This "break" indicates the beginning of acetone and butanol formation (Reilly *et al.* 1920). If it does not occur, butyric and acetic acids are the main products.

Failure to break under industrial conditions is not always understood; autoclaving glucose with yeast extract can have this result, and so can addition of base. Both these can also influence the growth of the bacteria, and some workers regard the break as merely marking the cessation of the growth phase (Dyr and Protiva, 1956).

Mechanism of the Fermentations

Source of Acetone: It was an early observation, made during the industrial acetone fermentations of World War I, that the yield of acetone can be increased by adding acetic acid (Reilly *et al.*, 1920). In more recent experiments with non-growing cell suspensions of *CL. acetobutylicum* BY, which convert about 22 per cent of the fermented glucose into butanol, the yield of acetone was raised from 4 to 16 per cent by adding acetate. Since 1 mole of acetate yields about per cent mole of acetone and 1 mole of CO_2, the process evidently goes via acetoacetate.

$$2\ CHaCOOH \longrightarrow CH_3COCH_2COOH \longrightarrow CH_3COCH_8 + CO_2$$

In confirmation:

(1) Added acetoacetate is rapidly and quantitatively converted to acetone and CO_2 (Johnson *et al.*, 1933);

(2) Isotopically labelled acetate yields labelled acetone (Wood *et al.*, 1945). Acetylco-enzyme A is an intermediate. Acetic acid alone is not converted to acetone, but simultaneous fermentation of glucose or other substrate has to go on (Janke and Siedler, 1937; Davies, 1942). This is because the condensation of acetic to acetoacetic acid, which goes via acetyl-co-enzyme A, requires energy (-AF = -16,000 cal).

The acetone-forming enzyme, acetoacetic decarboxylase, has been prepared from *CL. acetobutylicum* (Davies, 1943). Its *pH.* optimum is 4.5 to 5, as might be expected. It does not decarboxylate any other known keto acid. The reaction is probably due to combination of the ketone group with an amino group of the enzyme to form a Schiff base.

Source of Isoproponol

Isopropanol is almost certainly formed by reduction of acetone. As with acetone formation by *Cl. acetobutylicum* so with isopropanol formation by *Cl. butylicum,* addition of acetic acid increases the yield. One mole of acetic acid gives about per cent mole isopropanol. Added acetone also increases the yield of isopropanol, and with *Cl. toanum*, acetone appears first and then decreases while isopropanol is formed (Baba, 1953-1954). C^{14}-labelled acetone produces labelled isopropanol, leaving little doubt that the reduction goes easily.

Butyric Acid Formation

The appearance of butyric acid was first ascribed to the splitting of hexose into a C_4-f a C_2 fragment, but this is negatived by the old observation that some organisms can convert C_3 compounds to butyrate.

Calcium pyruvate is fermented by *Cl. butylicum* to acetate and butyrate (Brown *el al.*, 1938), and glycerol similarly by *Cl. acetobutylicum* (Simon, 1947). Following Neuberg's *Abfangverfahren* with yeast, the formation of acetaldehyde was demonstrated in the butyric fermentation (Neuberg and Arinstein, 1921). This led to a second theory, namely, that both butyric and acetoacetic acids were formed via acetaldehyde, which would be condensed to acetaldol. This could undergo an internal oxido-reduction to butyric acid, could be dehydrogenated to acetoacetic acid, or reduced to butanol. Unfortunately for this theory, neither acetaldehyde nor acetaldol yielded much butyric acid or butanol in the fermentation (Janke and Siedler, 1937). Furthermore, the observations have not been confirmed, and even the identity of Neuberg's culture is uncertain.

It is now clear that butyric acid is indeed formed from a C_2 compound, but that this is acetyl-co-enzyme A, formed from acetate or from pyruvate. The evidence for the intermediate role of acetate is threefold: First, the addition of labelled acetic acid, $C^{13}H_3COOH$, to active fermentations gives rise to labelled butyric acid (Wood *et al.*, 1945). Second, the caproic bacterium, *Cl. kluyveri*, forms butyric acid from acetic acid and alcohol:

$$CH_3COOH + C_2H_5OH \longrightarrow C_3H_7COOH + H_2O$$

If $CH_3C^{13}OOH$ is used, the resulting butyric acid has the isotope in both the carboxyl and α-carbon atoms (Barker *et al.*, 1945; Barker, 1947). This means that two molecules of the labelled acetic acid must be reductively condensed (the hydrogen being provided by the ethanol).

If dried *Cl. kluyveri* is extracted with dilute sodium sulphide solution, the resuiting extract converts a mixture of ethanol and acetate to butyrate. Labelled acetate gave rise to labelled butyrate (Stadtman and Barker, 1949a). In an atmosphere of hydrogen, butyrate was formed from a mixture of acetate and acetyl phosphate.

Butyrate is formed less readily from pyruvate, doubtless because of the need for extra hydrogen, which pyruvate does not supply. *Cl. butyricum* and *Cl. acetobutylicum* convert pyruvate only to acetate, CO_2, and H_2 (Kubowitz, 1934); but a cell-free extract of *Cl. butylicum* converts pyruvate plus phosphate into acetyl phosphate, CO_2, and H_2. This observation is critical because acetyl phosphate can come in to equilibrium (via phosphotransacetylase) with acetyl-co-enzyme A. The participation of CoA was first indicated by work with pigeon liver extracts, which showed that this co-enzyme is essential when acetate plus ATP brings about the acetylation of aromatic compounds, and especially by the isolation of acetyl-CoA from yeast and the evidence that it acts as an acetylating agent therein (Lynen *et al.*, 1951-1952; Lynen, 1954).

The existence of the phosphotransacetylase, which catalyses the equilibrium between acetyl phosphate and acetyl-CoA, was then discovered in *Cl. kluyveri* extracts. Thus pyruvate is converted to acetyl phosphate (and vice versa) via acetyl-co-enzyme A. But it is the next step that is most characteristic; two molecules of acetyl-CoA condense to form acetoacetyl-CoA (Lynen, 1954); this C_4 compound is the key to all the C_4-forming reactions of the Clostridia.

In liver this condensation requires CO_2 and goes via malonyl-CoA, but this probably does not occur in Clostridia. Acetoacetyl-CoA is now reduced. The reaction takes place in three steps, namely, (1) to hydroxybutyryl-CoA, (2) to crotonyl-CoA, (3) to butyryl-CoA (Stadtman, 1956; Barker, 1957).

Free butyric acid may be formed either by hydrolysis or by the transfer of co-enzyme A to acetate by the CoA-transphorase enzyme of Stadtman (1952), to form acetyl-CoA again; the latter would ensure the rapid recirculation of the co-enzyme. Three of the key enzymes required for these reactions have been shown in *Cl. acetobutylicum* and *Cl. sticklandii* (Stern *et al.*, 1956), as well as in *Cl. kluyveri*, although here the source of acetyl-CoA may be acetaldehyde. The two reductions make use of TPNH. The enzyme crotonase, has been crystallised from liver and shown to act on CoA esters of all the unsaturated acids with 4 to 9 carbon atoms (Stern *et al.*, 1956).

The formation of butyrate by condensation of two Co units explains the curious fermentations of lactate, which go only in presence of a second substance; with *Cl. lactoacetophilum* this is acetate, but with *Butyribacterium rettgeri* it is, surprisingly enough, CO_2. In the former case about 1 mole of acetate is used per mole of butyrate formed (Bhat and Barker, 1947-1948), and it is deduced that the lactate is converted to acetate, which is then condensed (via acetyl-CoA) and reduced to butyrate. Although some hydrogen is lost as gas, nevertheless, more remains than can be accepted by the 1 mole of acetate formed; hence more acetate must be added. The Butyribacterium evolves no hydrogen at all, the excess being somehow combined with CO_2 to form a second mole of acetate (Barker *et al.*, 1945).

If $C^{14}O_2$ was supplied, the acetate and butyrate were found to be labelled in all C atoms. Poly-hydroxy-butyric acid is a common bacterial polymer, found in bacilli, pseudomonads, Rhizobia, and in purple bacteria (Forsyth *et al.*, 1958). *Bar. megaterium* possesses an enzyme which causes the polymerisation very rapidly (Merrick and Doudoroff, 1961); some Clostridia may do the same. The butyric fermentation is inhibited by CO, and the inhibition is reversed by light (Kubowitz, 1934). Absence of iron has the same effect; in both cases only lactate is produced. Evidently, therefore, a hemin-containing enzyme is involved, and it perhaps functions at the initial split of pyruvate to acetyl-CoA or acetyl phosphate.

The enzyme is not the same as formic hydrogenlyase, since extracts of *Cl. butylicum* which evolve H_2 from pyruvate do not decompose formate (Koepsell *et al.*, 1944). It may be a related hemin-enzyme that operates with pyruvate instead.

Although the intermediate steps between glucose and pyruvate are presumably the same as in the alcoholic and lactic fermentations, it is an odd fact that neither hexose-l,6-diphosphate nor phosphoglycerate produced any butyric acid when incubated with fresh or acetone-dried *Cl. acetobutylicum* (Simon, 1947). Even glucose-6-phosphate gave only very small yields of butyrate. Probably the phosphate esters are poorly taken up.

Source of Butanol

The experiments on the influence of pH show that when butanol is not formed, butyric acid appears in its place. Addition of butyric acid to *Cl. acetobutylicum* fermentations, however, has given conflicting results. Simon and Weizmann (1937) got an increase in the oil, which may or may not have been butanol; Bernhauer and Kurschner (1935) in one experiment got a 90 per cent yield of butanol from butyric acid added to corn mash; Lerner (1941) found that about half the added butyrate was converted to acetate, the remainder being recovered unchanged. However, with *Cl. butylicum* reduction was clearly shown (Osburn *et al.*, 1938). That the added butyric acid gave rise to about per cent mole butanol; the rest was converted to acetic acid, CO_2, and a little isopropanol. Acetic acid did not yield butanol, but only isopropanol, i.e., the acetoacetic acid formed must have been decarboxylated. With *Cl. beijerinckii*, Gfeller (1954) similarly found that addition of butyric acid nearly doubled the yield of butanol and increased the acetone plus isopropanol by 35 per cent.

At the "break," when butanol and acetone begin to be formed, both acetate and butyrate in the mash decrease

(Peterson and Fred, 1932). The decrease of acetate is of course expected, since it is the source of acetone, or of isopropanol, but the decrease of butyrate points to its reduction to butanol. Fermenting with H_2 under pressure (20 atoms) decreases the yield of butyric acid and raises that of butanol (Eliasberg, 1930).

Finally, the reduction has been confirmed by isotope studies. Doubly labelled butyric acid, $CH_3C^{13}H_2CH_2C^{13}OOH$, was added to fermentations of corn meal, and 84 per cent of the C^{13} was recovered in the butanol (Wood, Brown, and Werkman, 1945). With *Cl acetobutylicum,* 2 per cent was recovered as acetone, and with *Cl. butylicum* 3 per cent as isopropañol.

These results seem conclusive. Also, $C^{13}H_3COOH$ was converted to butanol, more indeed (55 per cent) than to acetone (15 per cent) or CO_2 (18 per cent). This indicates that when the acetic acid has been converted to acetoacetyl-CoA, its subsequent reduction by these organisms is faster than its decarboxylation. With an extract from *Cl. kluyveri,* butyryl phosphate is readily reduced to butanol (Stadtman and Barker, 1950). Now the formation of butyrate already involves a reduction (acetoacetyl-CoA + 4[H]), so that when part of the acetoacetate is converted to acetone, the hydrogen acceptor necessary for the conversion of pyruvate to acetate is lost: it is for this reason that the extreme reduction of another acetoacetate molecule to butanol is essential. The formation of butanol could therefore be considered a necessary consequence of the formation of acetone:

Table: Increase in Yields of Products (as M per 100 M Glucose Fermented) When Butyric or Acetic Acids Are Added to Glucose Fermentation by Cl. butylicum

(For simplicity, only the increases resulting from the addition, and not the absolute amounts, are here shown.)

Added	*Butanol*	*Isopropanol*	*Butyric Acid*	*Acetic Acid*	*CO_2*	*H_2*
Butyric acid:						
38.5	19	8	-7	15	8	-27
41.5	19	5	7	23	12	-5
64.3	35	12	-5	23	26	

Contd...

Added	*Butanol*	*Isopropanol*	*Butyric Acid*	*Acetic Acid*	CO_2	H_2
Acetic acid:						
43.2	-5	30	-3	-4	18	-34
65.3	3	32	-6	0	1	

Caproic Fermentation

Barker, in 1935, discovered a related fermentation while studying the formation of methane. In this process a large anaerobe, named *Cl. kluyveri*, grew on a mixture of ethyl alcohol and acetate to form a mixture of butyric and caproic acids (Barker and Taha, 1942). The medium at first contained alcohol only, and the organism was always associated with a methane-producing form, but it developed that the role of the latter was only to produce acetic acid from the alcohol (Barker, 1947).

If sufficient acetate is added, *Cl. kluyveri* grows well in pure culture (requiring biotin and p-aminobenzoate as growth factors) and converts ethanol and acetate quantitatively to butyrate, caproate, and H_2. Sugars are not used, and no CO_2 is formed. The formation of butyrate has been described above: the alcohol is dehydrogenated, forming acetyl-CoA which condenses to acetoacetyl- CoA and is then reduced. The formation of caproate follows a similar pattern, a third molecule of acetyl-CoA condensing with the butyryl- CoA and the product being again reduced.

By analogy with animal cells, the condensing unit may actually be malonyl-CoA. The need for alcohol is as source of hydrogen for the reduction, although some of the hydrogen is lost as gas. Caproate is also formed in small yield by *Butyrib. rettgeri* from acetate and lactate, the lactate here evidently substituting for the ethanol.

The validity of reaction (6) is supported: (1) by a fermentation with butyrate- 1-C^{14} (butyric acid labelled in the carboxyl), the caproic acid then containing almost all the isotope; and (2) by experiments with cell-free extracts (Stadtman and

Barker, 1949a) which convert mixtures of ethanol and acetate to butyrate and caproate. If propionate was substituted for acetate in the medium (with live cells), then acetate and valerate were the main products, and some heptanoate also occurred (Bornstein and Barker, 1948):

$$C_2H_6OH + C_2H_6COOH \longrightarrow C_4H_9COOH + H_2O$$

$$C_2H_5OH + C_4H_9COOH \longrightarrow C_6H_{13}COOH + H_2O$$

Finally, the fermentation also involves a parallel to the butanol process, because in the presence of H_2 caproylphosphate is reduced by dried cells of *Cl. kluyveri* to small amounts of capryl alcohol (Stadtman and Barker, 1950). However, capryl alcohol is not formed in vivo, nor indeed is any alcohol higher than propanol.

Synthesis of Lipids in Bacteria

This is an appropriate point, therefore, to consider what is known of the synthesis of more complex lipids in bacteria. Most of this work, which is very recent, stems from Barker's studies of *Cl. kluyveri.*

Straight-chain Fatty Acids

The condensation of acetyl-CoA to chains of 4 or 6 carbon atoms might appear to be the basis for the formation of chains of any length. However, in animal tissues long-chain fatty acids are not formed in quite this way.

Instead, the acetyl-CoA is first carboxylated to form malonyl-CoA, and it is this that condenses with acetyl-CoA (Brady, 1958; Wakil and Ganguly, 1959; Lynen *et al.*, 1959):

> Biotin functions as the CO_2 carrier and may form N-carboxy-biotin as intermediate. The replacement of the new COOH by a second acetyl group is believed to occur after malonyl-CoA has been bound to the enzyme by an SH group on it (Lynen *et al.*, 1959; Brady, 1961).

The resulting acetoacetyl-CoA is reduced and can then accept a second malonyl-CoA to form a C-6 chain, and so on. Although the enzyme system that fixes CO_2 into malonyl-CoA, and exchanges it in presence of a higher fatty acid, is present in *Cl. kluyveri* (Vagelos and Alberts, 1960), still this system is not necessary for the formation of butyrate; acetyl-CoA alone will form butyrate, and malonyl-CoA does not even enter the molecule (Goldman *et al.*, 1961). The malonyl-CoA system therefore functions only for the long-chain acids; specifically, it works with caproyl-CoA and higher members of the group.

Among the bacterial acids are found a number of monounsaturated compounds. These are probably formed by oxidation of saturated acids, but in the Clostridia, where such oxidation could not occur, they are formed by loss of water from the hydroxy acids.

Branched-chain Acids

Steroids, carotenoids, and such branched units as those in co-enzyme Q are synthesised from a basic branched unit which is probably 5-phospho-mevalonic acid (Chaykin *et al.*, 1958).

The HMGCoA is reduced by TPNH to mevalonic acid, and CoA liberated, by an irreversible reaction which takes place while the substrates are bound to the enzyme (Popjak and Cornforth, 1960):

$$RCO\text{-}SCoA + 2\,TPNH + 2\,H^+ \longrightarrow RCH_2OH + CoA\text{-}SH + 2\,TPN+$$

The condensation of mevalonic acid via its phosphate and the related isopentenyl pyrophosphate involves loss of both phosphate and carboxyl, producing a highly reactive C_r,H_8 unit, which yields the long branched-chain isoprenoid structures with the basis.

The Mycobacteria, are notable for high- lipid content, and

their lipids contain some branched acids of a quite different type.

Phthienoic acid is present in both human and bovine strains of *Mycob. tuberculosis,* and is among the preparations that produce marked tubercles in mice; suggestively, it is absent (along with some other saturated acids) from the a virulent strains (Cason *et al.,* 1956). Mycocerosic acid:

$$CH_3(CH_2)_nCHCH_2CHCH_2CHCH_2CH\ COOH$$

| | | |

$$CH_3\ CH_3\ CH_3\ CH_3$$

is very similar. Lederer and co-workers (Lederer, 1961) have found labelled propionic acid to be incorporated entire into this acid, and it seems likely that the Mycobacteria condense *propenyl-CoA* in the same way as the Clostridia use acetyl-CoA.

After three or four propionic units have been thus condensed, malonyl- CoA units are added to give the remaining straight chain. The second branched type is the *mycolic acids,* with two or more long straight chains, e.g., corynomycolic acid, the simplest, isolated from *C. diphtheriae.*

This was proven by growing the cells on palmitic acid-l-C^{14}, when the corynomycolic acid contained C^{14} only in the COOH and CHOH groups (Gastambide-Odier and Lederer, 1960). The reaction is important, for these acids occur widely in the group: Corynebacteria contain the Cis acid Nocardias contain the CM acid *Mycob. avium* and *phlei* contain the CM acid *Mycob. tuberculosis* contain the *Cm* acid. The higher members probably have four branches, two of $C1_8$ and two of C_24 or C_{26} (Lederer, 1961). A mutant Mycobacterium actually needs $C_{26}H_{54}O_2$ for growth, so perhaps uses it to form a mycolic acid (Karlsson, 1956).

A third branched type bears a single CH_3 group, like 10-methyl- stearic acid, present in several Mycobacteria, or a cyclopropane group as in lactobacillic acid. These are probably formed by addition of a single formyl group to an unsaturated acid. C^{14}-formate is incorporated into lactobacillic acid by *L. arabinosus,* and the unsaturated acid, 11-A-octadecenoic, is converted to lactobacillic acid with high efficiency (Hofmann and Liu, 1960). The formyl transfer may be made by folic acid. Why should bacteria synthesise such a variety of specific lipids? Some are perhaps formed incidentally as by-products of another synthesis, just as, in a way, lactic or butyric acid is produced in the course of acquiring energy for growth. But most are doubtless essential for the cell's life.

Many of the cell's lipids are present in the membrane, which contains both true fats and phospholipids or phosphatidic acids (Asselineau and Lederer, 1960). Although the composition and structure of membranes are not yet understood, the fact that they control all the relations between the cell and its environment makes their properties critically important and specific. In gram-negative cells, lipids enter not only into the membrane but into the cell walls as well, e.g., Salmonella walls are 22 per cent lipid, and lipoprotein makes up some 80 per cent of the total wall weight. In a Myxococcus 50 per cent of the (trypsiri-treated) wall was lipid (Mason and Powelson, 1958). It is safe to predict that the study of the role of lipids in walls and membranes will extend greatly in the future.

Macerans and Polymyxa Fermentations

These processes have been comparatively little studied. In Schardinger's original study of *Aerobac. macerans* (1905), he found that 7 per cent of the potato starch used in the medium was converted to acetone and 21 per cent to ethanol.

Acetic and formic acids, but no butyric, were formed. *Aerobac. polymyxa* was regarded by Beijerinck (1894) as giving, from malt, mainly CO_2 and gum, with small quantities of hydrogen and traces of butanol, but later it was found to produce 2,3-butylene glycol in quantity, a 25 per cent yield from glucose (Donker, 1926) and a 27 per cent yield from cornstarch being recorded (Kooi, 1946).

Another 20 per cent is converted to ethanol, and 4 to 10 per cent to acetone. Thus, in general the fermentation is like that of Aerobacter, except for the acetone formation. Under the name of *Bacillus acetoelhylicum,* the *macerans* fermentation has been patented (Northrop, 1919). The yield of acetone is about 13 per cent when corn is used as substrate, and that of ethanol up to 30 per cent, or on glycerol up to over 40 per cent. The fermentation is greatly promoted by addition of iron as $Fe(OH)_3$ suspension (Roberts, 1947).

A recent portrait of *Bac. polymyxa* or *"asterosporus"* one of the Aerobacilli. Like *Aerobac. macerans,* it ferments practically all carbohydrates, including starch, inulin, pectin, and hemicellulose.

Anaerobes and Oxygen

Although the Aerobacilli are morphologically like the clostridia and often accompany them in butanol enrichments of grain mash, they are aerobic and die off when the fermentation is well advanced. Correspondingly, of course, the clostridia die off if conditions become aerobic, yet at least some of the enzymes of the clostridia can function under aerobic conditions.

Stadtman and Barker (1949b) found that an enzyme from *CI. kluyveri* can oxidise alcohol to acetal-dehyde and acetyl phosphate, and can oxidise butyrate to acetyl phosphate and acetate. Thus, strictly anaerobic organisms use, in fermentation, enzymes that can also function with oxygen. It should be noted, too, that the crotonase-thiolase-3-hydroxy- butyryl-CoA dehydrogenase group of enzymes functions also in animal tissues that are well supplied with oxygen. Dried cells of *Cl. kluyveri* can oxidise butyrate, and H_2O_2 accumulates (Lieberman and Barker, 1954), doubtless because of the participation of flavoprotein as H acceptor. Apparently the H_2O_2 inhibits the fission of acetoacetyl-CoA to acetate, but this in itself does not seem of fatal importance for growth.

If we recall that part at least of the fermentation, namely, the formation of acetone, can be carried out by an aerobe (*Aerobac. macerans*), the exact reason for the strictly anaerobic nature of the clostridia becomes difficult to give. The absence of the cytochrome system in these cells may well make oxygen inert for them, but would not explain why oxygen kills them. Beijerinck (1894) was convinced that the strict anaerobiosis of the butylic clostridia was a function of the high-sugar media usually used. On a poor medium, containing 1 per cent peptone and per cent per cent starch, at 10° to 12°C, he could obtain appreciable growth and butanol formation only under aerobic conditions. His conclusion, that the bacteria when growing in the natural media of soil and water need occasional contact with oxygen in order to maintain growth, suggests some sort of aerobic vitamin synthesis. This might repay investigation by modern methods.

Production of Riboflavin

The butylic fermentation liquor is bright yellow; it is rich in riboflavin, synthesised by the bacteria. This has turned out to be a most valuable by-product. Addition of unpolished rice increases the yield of riboflavin, while excess iron decreases it, since steel fermenting tanks give poor yields. Amino acid mixtures, especially arginine plus asparagine, as well as biotin, also increase the yield of riboflavin (Imai, 1956), which at its maximum may reach 0.3 mg per gram dry matter, or 300 ppm. The *polymyxa* fermentation also produces riboflavin, but in lesser amounts (Adams and Leslie, 1946). Some of the yeastlike Ascomycetes, including *Eremothecium ashbyi* (Guilliermond, 1936), *Ashbya gossypii*, and several Candida species also form much riboflavin from sugars and compete with the clostridia in industrial importance. Iron and zinc are both critical, the optimum Fe level for Candida being about 0.01 ppm; about 5 mg of iron per litre inhibits riboflavin formation drastically (Goodwin and McEvoy, 1959). For Candida, Zn is essential, and Fe acts only on growth (Kniisel, 1957).

With Eremothecium growing on a synthetic medium,

thiamine was the main factor promoting riboflavin formation, although it did not promote growth (Osman and Kamel, 1960). *Aspergillus niger* can form riboflavin also, and again the process is separate from the growth, because the addition of sugars like sucrose and arabinose, or the use of nitrate as sole N source, or limitation of the Mg supply, or Hg salts, all retard growth but increase the riboflavin yield (Sarasin, 1953). Numerous additions of C^{14}-labelled substrates, followed by breakdown of the riboflavin and location of the isotope, have led to the recognition that ring A is formed from acetyl groups or acetoin, while ring C is derived from purines, especially adenine; the ribityl group comes from glucose (McNutt, 1954; Goodwin and Jones, 1956; Brown *et al.*, 1958).

Acetoin from Adenine

However, adding these substrates in quantity to Ashbya or Eremothecium cultures increases the riboflavin very little. Adenine and serine do increase the yield somewhat. From both Aerobacter and Eremothecium an enzyme system that catalyses the final steps in riboflavin synthesis from precursors, including acetoin, has been extracted (Katagiri *et al.*, 1959).

Inorganic Hydrogen Acceptors

Sulphate Reduction: Any reasonably sensitive nose can detect that H_2S is often produced in nature; besides the special case of sulphur springs, H_2S can be noted above sewage, certain kinds of swamps and mud, and saltwater pools. Some of this may come from proteins, as in the case of rotten eggs; H_2S is readily liberated from cysteine.

However, the H_2S that is usually detected where vegetation is decaying comes from another source. Beijerinck, doubtless stimulated by the odour of the Delft canals in summer, made a classic investigation of this (1895). Enrichment cultures made from ditch mud in a medium containing lactate as organic nutrient, with inorganic salts and 2 per cent $CaSO_4$, yielded H_2S on incubation. The organism responsible, a short spirillum or vibrio, was a strict anaerobe and was only obtained by making mixed cultures together with a facultative anaerobe (Aerobacter) which consumed the last traces of oxygen in the medium. Later, pure cultures were obtained by including enough H_2S to react with the residual oxygen (van Delden, 1904). The bacteria, subsequently named Desulfovibrio, were producing H_2S by *reduction of sulphate:*

$$H_2SO_4 + 8(H) \longrightarrow 4\ H_2O + H_2S$$

The organisms are curved rods and gram-negative typical vibrios, with a single polar flagellum (or occasionally two). In addition a straight rod, *Clostridium nigrificans* has the same physiology (Campbell *et al.*, 1957; Stiiven, 1960). The action of this organism has been thought to explain the occasional appearance of spores in sulphate-reducing cultures, especially at temperatures above 50°C, and by an odd chance the spores are vibrio-shaped (Stuven, 1960). However, another true vibrio, *D. orientis,* has been reported to reduce sulphate and to form spores (Adams and Postgate, 1959); the spores are said to occur rarely but "on three occasions we have observed *ca.* 90 per cent sporulation".

The sulphate-reduction process explains the frequent association of H_2S in nature with salt waters. Most such waters contain sulphate, and decaying vegetation in them provides organic matter and establishes anaerobic conditions. The H_2S often combines with iron salts to precipitate black FeS; hence, H_2S is associated with "black mud." The Black Sea is named for this, and the association is common everywhere in marine inlets; the shores of the Great Salt Lake provide another example.

The peculiar conditions prevailing in oil fields, where quantities of organic matter are found at great depths, favour the growth of the anaerobic sulphate-reducers. A connection between H_2S formation and petroleum deposits was pointed out by Murray Stewart in 1923 and by Bastin. Since many oil fields are near the sea, which contains sulphate, the relationship is readily under- [2] The media used by different workers are very similar (Starkey, 1948). Van Delden's, the simplest, is, in grams per litre: Na lactate, 5; asparagine, 1.9; K_2HPO_4, 0.5; $MgSO_4.7H_2O$, 1.0; $FeSO_4$, a trace. By using ammonium nitrogen the growth of sporeformers is discouraged. Elion's "first-rate culture medium" (1924) was: Na lactate, 3; NH_4Cl 1.0; K_2HPO_4, 0.5; $MgSO_4.7H_2O$, 2.0; $CaSO_4$, 1.0; ferrous amm. sulphate, 0.5, or metallic iron.

A similar medium with one small iron nail added per tube has been used for counting sulphate-reducers in natural waters (Abd-el-Malik and Rizk, 1958). Oxidation of the iron helps to establish anaerobic conditions and allows the growth of sulphate-reducers in tubes open to the air, stood, although it has not yet been shown that paraffin hydrocarbons can serve as hydrogen donors for the reduction.

There are other sites besides oil fields where sulphate-reducers have played a major part in geology. In some German salt mines the carnallite ($KCl\text{-}MgCl_2 . 6H_2O$) is blackened with flecks of FeS; elsewhere it contains red ferric oxide and residues of the highly aerobic iron bacteria, so that H_2S must have permeated the salt after deposition, suggesting the activity of a very halophilic Desulfovibrio (Miiller and Schwartz, 1953). Elemental sulphur is often found in rock deposits, where it has been ascribed to H_2S produced bacterially from sulphate; since the H_2S could not have been subsequently oxidised in the anaerobic deposits, it probably reacted chemically with sulphate or SO_2 to deposit sulphur.

Many microbes, and all algae and green plants, must be able to reduce sulphate in small quantities because they grow and form protein when sulphate is their sole source of sulphur. In the ascomycete *Ophiostoma multiannulatum* this ability is genetically controlled, and mutants can be readily obtained that cannot reduce sulphate, although these can still use cysteine, cystine, or sulphide for growth (Fries, 1946). At least three genes control sulphate-reduction, and some of the mutants can be adapted back to growth with sulphate. Such reduction of sulphate *for growth* is to be clearly distinguished from its reduction *in bulk* with evolution of H_2S. Perhaps the former process, followed by autolysis, explains the aerobic evolution of H_2S by *Bac. megaterium* growing on sulphate media (Bromfield, 1953). As organic compounds (i.e., hydrogen donors) for Desulfovibrio, a variety of substances will serve, including lactate, ethyl, propyl or butyl alcohol, glycols, some amino acids, and glucose (Baars, 1930). With lactate or alcohol, acetic acid is the main product, and the data fit well with the following equations (Baars):

$$2\,CH_3CHOHCOOH + H_2SO_4 \longrightarrow 2\,CH_3COOH + 2\,CO_2 + H_2S + 2\,H_2O$$

$$2\,CH_3CH_2OH + H_2SO_4 \longrightarrow 2\,CH_3COOH + H_2S + 2\,H_2O$$

In terms of equation, each molecule of these substances acts as a source of 4(H), e.g.:

$$CH_3CH_2OH + H_2O \longrightarrow CH_3COOH + 4(H)$$

Butyrate goes almost completely to CO_2, with a little acetate, i.e.,

$$C_3H_7COOH + 6\,H_2O \longrightarrow 4\,CO_2 + 20(H)$$

also be oxidised by hydroxylamine, nitrite, and even O_2, but the significance of this is doubtful, in so strict an anaerobe as Desulfovibrio. *D. orientis* has a different cytochrome, with band at 557 m. These phenomena strikingly parallel nitrate reduction, where also the nitrate acts as hydrogen acceptor by way of a cytochrome. *M. denitrificans* reducing nitrate with hydrogen gas also compares with the autotrophic growth of Desulfovibrio. The first product of sulphate-reduction is probably adenosine-5-phosphosulphite, since cell-free preparations reduce SO_4 to SO_3 by H_2 only in presence of ATP (Peck, 1959), 1 mole of ATP being used, and one pyrophosphate split off, for each H_2. TPNH can also function as reducing agent.

Further reduction can occur all the way to sulphide

(Ishimoto, 1959). The reaction between ATP and sulphate, yielding adenosine-5-phosphosulfite, has been studied in yeast (Robbins and Lipmann, 1956; Wilson and Bandurski, 1958; Hilz and Kittler, 1958, 1960); it is specific for ATP but not for sulphate, and occurs also with selenate, sulphite, chromate, molybdate, and tungstate. Since all these except sulphite inhibit sulphate-reduction, they probably do so by competing for ATP. Thus the combination with adenosine evidently makes the sulphate group more easily reducible. The sulphate reductase appears to contain lipoic acid.

Among other, less drastic, reductions, that of tetrathionate to thiosulphate is interesting (Knox and Pollock, 1944-1950): $S_4O_6 + 2\ e \longrightarrow 2S_2O_3$. This reactiobn, due to the enzyme tetrathionase, is brought about by Salmonella, in which it is readily induced. It is the reverse of the conventional iodine titration of thiosulphate; with its aid, the tetrathionate is made available for growth.

Since H_2S is continually being produced, it must also be removed. This is done by oxidation, supporting the growth of both chemosynthetic and photosynthetic autotrophs. Many minor sulphur compounds are also involved, especially in soil (Starkey, 1950). The resulting "sulphur cycle" is comparable, though not in complexity, with the nitrogen cycle (Gregory and Robbins, 1960).

Methane Formation

Methane is readily formed when plant material decays in waterlogged anaerobic soil. A flickering blue light, seen over swamps, was formerly said to have lured travellers deeper and deeper into the bog and so to their doom. This *ignis fatuus* or false fire is due to the ignition of the methane, probably by traces of phosphorus hydrides.

Methane is produced in such large quantities in the anaerobic decomposition of sewage that the gases from the Imhoff tank are sometimes piped off for heating or lighting purposes. It was Sohngen (1906, 1910) who first made enrichment cultures that fermented the lower fatty acids to methane. No other hydrocarbon resulted.

A large sarcina was the most prominent organism. Later, a thermophilic bacterium was found to produce methane from fatty acids or glucose, while crude cultures from faeces would attack cabbage and cellulose in the same way (Coolhaas, 1928). Other fermentable substrate are methanol, ethanol, butanol, formate, butyrate, formaldehyde, and acetone.

The fermentation goes best when *n* is an even number (Sohngen, 1910). In this reaction, the fatty acids are not decarboxylated, because no matter what the length of the carbon chain, only methane is produced; higher hydrocarbons have never been obtained in appreciable quantities (Thayer, 1931). The absence of higher hydrocarbons means that compounds of two or more carbon atoms are not being reduced; the corollary is that methane arises because a compound of one carbon atom, namely CO_2, is being reduced.

This brilliant suggestion made by van Niel in 1933 to Barker has been amply confirmed (Barker, 1936- 1943). It was, indeed, earlier observed that certain rods will form CH_4 from hydrogen and CO_2 (Sohngen, 1906; Fischer *et al.*, 1932; Stephenson and Stickland, 1933): $CO_2 + 4\ H_2 \longrightarrow CH_4 + 2\ H_2O$ In Barker's studies, pure cultures were obtained for the first time by using Na_2S to remove the last traces of oxygen, Minute colonies, made evident by gas production in deep agar tubes, yielded pure cultures of four types:

1. A large sarcina, evidently Sohngen's organism, called *Methanosarcina (Ms.) methanica*, and gramvariable, unlike other *Sarcinae;* this organism had also been studied by Smit (1930;).
2. A coccus, also gram-variable, called *Methanococcus (Me.)*.
3. A gram-negative rod forming long threads, called *Methano-bacterium (Mb.) Sohngenii;*

4. Another rod, forming "sporelike bodies" of low heat resistance and called *Methanobacterium (Mb.) Omelianskii.* This too is mostly gram-negative.

The first three ferment acetic and butyric acids, the last ferments alcohols. Two other rods and another Sarcina were obtained later (Kluyver and Schnellen, 1947; Mylroie and Hungate, 1954), and the methaneforming bacterium of cattle rumen has since been added (Carroll and Hungate, 1955); this last is a gram-positive non-motile rod, *Mb. ruminantium,* which can use H_2 or, more weakly, HCOOH as reducing agent. With these organisms the following facts have been established:

1. Ethyl alcohol and CO_2 are converted quantitatively to acetic acid and methane:

   $$2\ CH_3CH_2OH + CO_2 \longrightarrow 2\ CH_3COOH + CH_4$$

   Note the similarity to equation (3) of the *Desulfovibrio* above.

2. Butanol is similarly converted to butyric acid and methane:

   $$2\ C_3H_7CH_2OH + CO_2 \longrightarrow 2\ C_3H_7COOH + CH_4$$

3. In cooperation with *Cl. kluyveri,* ethanol can be converted, via acetic acid, to butyric and caproic acids. This is, of course, a side reaction.

4. $C^{14}O_2$ is reduced to $C^{14}H_4$ by *Mb. Omelianskii,* fermenting alcohol (Barker *et al.*, 1940; Stadtman and Barker, 1949), and by *Mb. ruminantium,* using CO_2 and H_2 (Smith and Hungate, 1958). In the former case, the cells had some radioactivity, indicating that 1.5 per cent of the CO_2 was assimilated; the inactive carbon of ethanol was also assimilated (Barker and Beck, 1941).

5. Even with methanol, using *Ms. methanica,* $C^{14}O_2$ gave radioactive methane. This is of special interest because the equation for the overall process,

   $$4\ CH_3OH \longrightarrow CO_2 + 3\ CH_4 + 2\ H_2O \quad (7)$$

could be interpreted as a reduction of methanol rather than of CO_2. Evidently the real reaction has two phases:

(a) $CH_3OH + H_2O \longrightarrow CO_2 + 6(H)$

(b) $CO_2 + 8(H) \longrightarrow CH_4 + 2\ H_2O$

Equation (7) represents four of (a) and three of (b).

6. Hydrogen gas can be used by *Mb. Omelianskii, Mb. ruminantium, Mb. formicicum,* and another sarcina, *Ms. Barkeri* (Kluyver and Schnellen, 1947; Smith and Hungate, 1958).

7. A striking variant is the use of CO. A mixture of CO and H_2 passed over a natural sludge gave rise to CH_4 and CO_2 (Fischer *et al.*, 1931,1932). By varying the gas concentration, it could be shown (with enrichment cultures) that the reaction occurs in two steps:

   $CO + H_2O \longrightarrow CO_2 + H_2$

   $CO_2 + 4\ H_2 \longrightarrow CH_4 + 2\ H_2O$

   With pure cultures of *Ms. Barkeri,* the same thing was shown by analysing the mixture before the reaction was over (Kluyver and Schnellen, 1947). *Mb. formicicum,* which ferments formate, similarly attacks CO and hydrogen (Mylroie and Hungate, 1954). Methane can even be produced by *Ms. Barkeri* from pure CO alone. Here the first of the two reactions is the only source of hydrogen, and the overall equation would be:

   $4\ CO + 2\ H_2O \longrightarrow 3\ CO_2 + CH_4$

8. Only in the fermentations of acetate and benzoate is methane formed from other sources than CO_2. With acetate (using enrichment cultures of Methanococcus) the methane comes mostly from the methyl group and the CO_2 from the carboxyl group, although towards the end a little CO_2 is converted to methane (Stadtman and Barker, 1949). This anomaly may mean that both CO_2 and acetate are converted into methane only via

an intermediate compound that is common to both. The direct decarboxylation of acetate seems open to doubt. With benzoate (using enrichment cultures from sewage sludge) high yields of CH_4 and CO_2 were obtained, but very little of the CO, added was reduced to methane (Clark and Fina, 1952). Apparently most of the methane came directly from benzoic acid. With the exception of these last two special cases, the data all show that the fundamental reaction in the methane fermentation is:

$$CO_2 + 4\ H_2A \longrightarrow CH_4 + 4\ A + 2\ H_2O$$

In the case of hydrogen, A of course is missing. This representation of CO_2 reduction shows great similarity to the reactions of photosynthesis. But whereas in photosynthesis, the CO_2 reducing enzymes are rather well understood, the methane-forming enzymes remain totally unknown.

Acetic Acid Formation from CO_2

A series of bacterial processes differing only moderately from the methane fermentation will be described quite briefly. In studying the conversion of CO and CO_2 to methane, Fischer *et al.* (1932) found that in certain cultures acetic acid is formed rather than methane, and the reaction approximates:

$$2\ CO_2 + 4\ H_2 \longrightarrow CH_3COOH + 2\ H_2O$$

Later a *Clostridium, Cl. aceticum,* was obtained in pure culture that carried out reaction (9) alone (Wieringa, 1936). Evidently this represents an alternative form of the generalised reaction (8), with acetic acid replacing methane. It appears now that this is a widespread bacterial reaction type, and four instances have been elucidated by Barker and his students.

1. The butyric fermentation due to *Butyribacterium rettgeri* differs from that caused by the butyric clostridia in that lactate is readily fermented. The yield of acetic and butyric acids is relatively high and that of CO_2 is low.

Accordingly Barker *et al.* (1945) considered that some of the CO_2 might be converted to fatty acids. The fermentation was allowed to proceed in presence of $C^{14}O_2$; as a result about 25 per cent of the C^{11} was recovered as acetic acid and about 50 per cent as butyric:

$$CH_3CHOHCOOH + H_2O \longrightarrow CH_3COOH + CO_2 + 4(H)$$

Then:

$$2\ CO_2 + 8(H) \longrightarrow CH_3COOH + 2\ H_2O$$

Other purines give similar reactions. When the fermentation was carried out in the presence of $C^{14}O_2$, both the acetic acid and the non-volatile material contained the isotope (Barker *et al.*, 1940). Evidently, therefore, CO_2 is being reduced to acetic acid, as in equation (10), as well as other compounds. *Cl. cylindrosporum* gives a similar fermentation which produces glycine in addition, and here $C^{14}O_2$ yields isotope in the CH_3 of acetic acid and the COOH of glycine (Barker and Eldsen, 1947). (3) Perhaps the most remarkable example is the fermentation of sugars carried out by *Cl. thermoaceticum* (Barker, 1944; Barker and Kamen, 1945). Glucose or xylose is converted almost completely to acetic acid, anaerobically, by this organism:

$$C_6H_{12}O_6 \longrightarrow 3\ C_2H_4O_2$$

Added $C^{14}O_2$ is readily recovered in the acetic acid. Hence the sugar must first be converted to acetic acid and CO_2:

$$C_6H_{12}O_6 + 2\ H_2O \longrightarrow 2\ CH_3COOH + 2\ CO_2 + 8(H)$$

and the resulting CO_2 then reduced as in equation. Pyruvic acid is probably an intermediate, since it is also fermented to acetic acid and CO_2.

Aerobic and Anaerobic CO_2 Reduction

All the cases surveyed above involve strict anaerobes and consist of the transfer of either hydrogen gas or the hydrogen from organic compounds to the acceptors, sulphate or carbon dioxide. When the acceptor is CO_2, the organism has at least the potentiality of growing autotrophically. The most primitive organisms on earth were probably anaerobes. When oxygen became available, it would have been preferred, as a hydrogen acceptor, to any other compound, organic or inorganic, because the combination with oxygen is more strongly exergonic than any alternative.

Correspondingly, the aerobic organisms are now the dominant forms. Although CO_2 still has to be reduced, we find the aerobes linking the reduction to an oxidation that provides the energy for it. Thus, *Hydrogenomonas,* which reduces CO_2 with the oxidation of hydrogen, will not combine CO_2 with hydrogen anaerobically (Kluyver and Manten, 1942). Yet CO_2 and EL are combined by the anaerobes described above. It follows that the linkage to an oxidation process in the aerobic forms (if one dare assume the direction of evolution) has become obligate over the years.

Growth and Protoplasm Formation

Normal Growth Curve: Under ideal conditions a bacterial cell can divide in two every 15 to 20 minutes. At this rate any reasonably sized culture flask would be filled to overflowing in about 12 hours. If the rate continued undiminished for 36 hours, the progeny of a single cell would cover the whole earth with a layer of bacteria a foot deep. Needless to say, what actually happens is that the growth rate changes rapidly with time. The study and interpretation of these changes are, therefore, of the highest importance. If a few bacteria are inoculated into a new medium, and small samples taken out from time to time for counting, by the conventional method of pouring plates and enumerating the colonies, there is obtained all or part of a curve. Such curves have been divided into eight phases:

1. The *stationary phase,* in which no multiplication occurs.
2. The phase of increasing growth rate. These first two phases together constitute the *lag phase.*
3. The *"logarithmic"* or *exponential phase,* in which the log of the number of cells increases linearly with time. This

means that each cell is dividing in two at a fixed time interval, i.e., the growth rate is constant and maximal.

4. A phase of decreasing growth rate, with many cells dying off.
5. A brief stationary phase, in which growth and death are just balanced, and the population is at the maximum the medium can support.
6. The phase of increasing death rate.
7. The "logarithmic" death phase; the inverse of phase 3. This is seldom studied at medium temperature and is mainly of interest in connection with heat sterilisation.
8. A phase, which may last a very long time, of decreasing death rate. Provided the medium does not dry up, a few survivors may be present almost indefinitely.

In interpreting the curve, it may be objected that the plating method does not actually count the bacteria, but counts colonies derived from them; i.e., it counts only those bacteria that could multiply to form colonies. Indeed, it is called the *viable count.* This difficulty can be avoided by actually counting all cells present (total count), which is done by pipetting a small volume of culture into a measured depression on a microscope slide and counting the visible cells, either under dark-field illumination or after staining. Generally these counts give results only a little higher than the viable count. The difference, in a careful study with *S. choleraesuis,* was about 25 per cent (Wilson, 1922). Hence the phenomena of the growth curve really do represent rates of change of the population. Much work has been done to explain the different phases.

Lag Phase

The first and most obvious problem is: Why do not the organisms begin to multiply immediately. Why need there be any lag Phases 1 and 2, the so-called "lag period," are not really to be expected a priori, considering that the medium is new. As a matter of fact, if we take for the inoculum cells from the actively dividing period, phase 3, they do not show any lag (Chesney, 1916). The lag is shown by inocula taken from phase 1 and also from phases 5 and 6. It is therefore, a function of the state of the cells when transferred; non-growing cells have to get into a growing or "embryonic" state before they can start dividing.

The first clue to the nature of this growing state was obtained by carefully observing the cells. It then appeared that during the lag they grow to much greater size than normal, then decrease again during the logarithmic phase. Henrici's data for *Bac. megatherium*. Two cycles are given, a subculture being made at eight hours when the cells have come down to the initial size. The actual size reached depends some- what on the concentration of the nutrient medium: when the medium was reduced to a quarter of its original strength, the maximum size fell from 8 /x to 6 /x (Henrici, 1923;).

Since the cells grow larger instead of dividing, there is no lag in the synthesis of protoplasm. If we follow, instead of cell numbers, the total protein, the nucleic acid, or the cell phosphorus, we see that the increase of cell material begins immediately. The total nitrogen, for instance, increases linearly, on a logarithmic scale, from the start (Hershey, 1938). Synchronised cultures show these relationships best. It is evident that many constituents must be increasing on a per cell basis during the lag phase.

The oxygen uptake per cell also passes through a sharp maximum, which is reached at about one and one-half hours with *E. coli*, i.e., after the end of the lag phase, and corresponds exactly to the time when the cells are at their largest. Again, if calculated per unit of bacterial nitrogen, there is no maximum, but the rate remains constant, i.e., oxygen uptake is a function of the amount of protoplasm and not of the number of cells. However, respiration is increased by growth. All these data, therefore, show that in young cultures the cells increase at once in size and in materials, but without dividing; it is only the division process that shows a lag. The lag phase should be called the "phase of cell enlargement" (Hershey, 1938).

The reason for the lag in cell division is not known, and a special cell division factor has been postulated (Lodge and Hinshelwood, 1943). Of its nature there are conflicting indications.

Apparently Mg ions promote cell division in *Cl. welchii*, and the cultures become filamentous without Mg (Webb, 1948), but it is hard to see how this could play a part in the "lag" phenomenon.

In yeasts, cysteine and glutathione promote cell division, while penicillin and also cobalt inhibit it, producing long "mycelial" growths (Nickerson and van Rij, 1949). Thus, if there really is a division factor in these cells, it might be a sulphhydryl compound. (Cobalt readily forms complexes with sulphhydryl). Many bacteria growing in the presence of penicillin also form filamentous growths because this antibiotic inhibits the formation of the cell wall. In Bacilli the lag can be reduced or prevented by adding glucose autoclaved with phosphate (Sargeant *et al.*, 1957).

The effective compound is a hydroxyacid, and several ordinary hydroxyacids have the same effect; possibly they act by chelating a metal ion. In Aerobacter, the lag is removed by adding glutamic acid in concentrations of 5 to 10 parts per million. The effect is not specific, since some other amino acids, and also succinate, have a similar effect (Dagley *et al.*, 1949). (The bacteria can form glutamate from succinate, however.) This suggests that the lag is concerned with amino acid formation. In some lactobacilli, the lag is reduced by adding extra vitamins. Probably cell division is complexly interrelated with nitrogen metabolism. During the lag phase the cells are less sensitive to unfavourable conditions than later. Thus, cells of *E. coli* in the first hour after transfer from a one-week-old culture were resistant to a 5 per cent NaCl (= 0.9 M), but similar exposure a few hours later caused great mortality. The sensitivity began to increase before there was appreciable cell division (Sherman and Albus, 1923-1924). The resistance to heat treatment is also increased at first, then rapidly decreases as cell division begins (Elliker and Frazier, 1938). Part of this resistance of the non-dividing cells may be simply due to their large size and consequent smaller relative surface.

Synchronised Cell Division

The time of emergence from the lag phase is evidently distributed statistically through the population. In recent years, several means have been developed to minimise distribution and start all cells dividing at once. One method depends on sudden changes in temperature. Many cultures, if chilled to 0°C in a nutrient medium and rapidly warmed, at once begin to divide. When Salmonella cultures are quickly raised from 25° to 37°C by adding a known volume of hot medium, the two deeply staining "nuclear" bodies each divide within about six minutes; within 20 minutes all the cells have divided once (Lark and Maal, 1954). This *synchronised division* can be repeated several times by cooling and reheating; the precise time intervals depend on the culture density and its growth phase. A method developed by Japanese workers depends on the observation that the smaller cells divide more rapidly than the larger ones. In *E. coli* strain B, synchronised division can be achieved by centrifuging into fractions of different cell size; on then being placed in fresh medium the small cells all divide in 30 to 40 minutes (at 37°C), while the large ones undergo two or three division cycles at 50-minute intervals (Maruyama and Yanagita, 1956). The separation can be made neatly by running the suspension through many layers of filter paper; the smallest cells come through first. While the cell numbers then go up in discrete steps, the rate of incorporation of leucine into protein, the rates of synthesis of DNA and of galactosidase, as well as the total amounts of DNA, RNA, and protein, all remain smoothly linear (Abbo and Pardee, 1960). It is evident that syntheses in the cell go on independently of cell division. A third method depends on the suppression of cell division exerted by very high cell densities. In *E. coli*, all divisions cease at 5.10^9 cells per millilitre (Yanagita and Kaneko, 1961). Syntheses still go on, as noted above, DNA and protein increasing 40 to 50 per cent in 50 minutes at 37° C. Then, when the suspension is suddenly diluted 1:100, all the cells divide.

Aeration

With the obvious exception of the anaerobes and microaerophiles, the growth of all microorganisms studied is highly dependent on oxygen.

This is true for yeast, ciliate protozoa (Phelps, 1936), and numerous bacteria (Monod, 1942; Rahn and Richardson, 1940-1942; van Niel, 1949). It is also true for fungi and higher plants. The curves show an enormous increase of growth with aeration. The decrease of growth and of sugar consumption at excessively high aeration rates may be due to the cooling effect of the air. For bacteria an example can be taken from the work of Winslow's group with *E. coli.* Here the addition of glucose raises the number of cells somewhat, but the effect of aeration is far greater. In peptone, the numbers increase only about three and one-half times when nitrogen is bubbled, but about 41 times with air. It may be noted, too, from the last column, that aeration decreases the output of CO_2 from glucose (Pasteur effect).

Not only does aeration increase the population density, but rapidly growing cells have a higher oxygen consumption than resting cells. Pirt (1957), for instance, found with Aerobacter in continuous culture that the respiration rate, per milligram of dry weight, varied linearly with the growth rate. Thus, the oxygen demand in growing cultures is increased even more, and to maintain maximum growth rate in a dense culture becomes very difficult.

Temperature

The Q_{10} of cell multiplication is commonly 2 to 3, but it becomes very high at around 0°C. Above the optimum temperature, the growth rate falls off very sharply, due to the excessively high Q_{10} of processes like enzyme destruction and protein denaturation. Temperature has very little influence on the total crop when other conditions are constant. With *E. coli* on a glucose medium, the total crop was almost constant from 23° to 33°C (Monod, 1942), while the generation time, g, would be about halved by the same 10°C rise. At much higher temperatures, the yield does, of course, fall off, and very sharply.

C. Nutrients

Two types of control by nutrients must be considered. In the first place, there are *specific* nutrients, essential for growth, like vitamins or, for some organisms, amino acids. Evidently one nutrient can easily be the limiting factor in an otherwise complete medium. This is the basis for microbiological assay of such compounds, in which the procedure is usually to determine the total crop after a fixed time. The relationship between yield and concentration of the factor is roughly linear over a certain concentration range and eventually flattens when some other factor becomes limiting. For each determination, an arbitrary "calibration curve" has to be constructed for the particular organism and the conditions of growth used. Sometimes, the shape of this calibration curve changes very materially with time.

This can be seen in a three-dimensional diagram like that for thiamine in the growth of Phycomyces. In a different category from growth factors are the *general* nutrients like the sugars, which provide both the energy and the carbon for synthesis. With these, organisms whose growth is not limited by specific growth factors can also show a response over a considerable concentration range. The total yield of *E. coli* and *Bac. subtilis* (both of which synthesise all required vitamins) varies linearly with sugar concentration up to about M/1000. It is striking that two different bacteria, utilising two different sugars, give the same straight line. Aerobacter shows similar variation in yield with concentration of medium (Dagley and Hinshelwood, 1938).

Although the total yield varies directly with the concentration of nutrient, the growth *rate* shows proportional response only at the lowest concentrations. Thus, the growth rate of *E. coli* increases up to a glucose concentration of about 40 mg per litre (M/4500) and then is absolutely constant (Monod, 1942); with strains that require tryptophan, the growth rate is proportional to tryptophan only up to 1 to 5 micrograms per litre, or about 10^{-8} M (Novick and Szilard, 1950).

With phosphate, the independence of growth rate and concentration is striking; the generation time for *E. coli* on a synthetic medium was constant over the range from 0.1 to 50 mg phosphate per litre, i.e., over a range of 500 times (Caldwell and Hinshelwood, 1951). The relation between growth rate and the concentration of a limiting factor has been utilised in an ingenious way to obtain a stationary population (Novick and Szilard, 1950). Nutrient medium containing a very low concentration of tryptophan is supplied continuously to a culture vessel from which the excess is allowed to overflow with a constant level device.

The culture, a tryptophan-requiring strain of *E. coli* increases in density and thus withdraws tryptophan from solution, which in turn reduces growth. Finally, a state is reached when the bacteria just divide once in the time it takes to renew the whole contents of the culture vessel, i.e., the population becomes constant. If the concentration of tryptophan (or other limiting factor) is *a* in the inflow tube and is reduced to *c* in the growth vessel.

For tryptophan Q is only about six million molecules, for glucose about 10^4 times greater. Any essential nutrient can be used as limiting factor in the same way. If a mutant should appear that uses the limiting factor more efficiently, i.e., can grow faster on the same concentration, it will replace the original form almost completely. Even a mutant that grows at the same rate as the original strain will be present in steadily increasing numbers, while one that grows more slowly will soon reach a constant number (Northrop and Kunitz, 1957). Thus, the apparatus is useful for the study of populations and evolution. Monod (1950) and Herbert *et al.* (1956) have used similar systems; in the latter case, with *Aerobacter cloacae*, mutations growing faster than the original strain were not encountered. An alternative control method is to maintain constant optical density with a photocell and adjust the flow rate accordingly, but this suffers from practical difficulties due to growth of cells on the walls or the optical window (Novick, 1955). These methods are increasingly important for large-scale cultures. As noted above, the cell size is about constant during the logarithmic phase, but its absolute size varies with the medium. In *Salmonella typhimurium* the influence of the medium on cell size is much greater than that of the temperature; the smaller cells have less RNA, less DNA, and a smaller average number of nuclei (Schaechter *et al.*, 1958). In spite of the large differences, the rate of protein synthesis on the different media was calculated to be the same per unit of RNA. When the cells were transferred to a richer medium, the RNA synthesis increased at once, but the DNA and the cell division rate increased only after 70 minutes (at 37°C). This delay is evidently analogous to the lag phenomenon.

Diauxic Growth Curves

A curious type of growth curve was described by Monod (1942, 1947; van Niel, 1949). *B. subtilis* growing on a synthetic medium, with nitrogen and phosphorus in excess, gives a normal growth curve, and the total yield is linearly proportional to the carbohydrate supply. However, when two carbohydrates are present in the medium, an entirely different growth curve may be produced, with two successive growth cycles separated either by a lag or by a slight decrease. Study of this phenomenon, "diauxic growth," showed that it occurred only with certain sugars. When the two carbohydrates were glucose, mannose, fructose, sucrose, or mannitol, any pair of these gave a normal curve. When one of them belonged to this group and the other was maltose, arabinose, sorbitol, or inositol, then the two-peaked or diauxic curve was obtained. In such diauxic curves, the first cycle corresponds to utilisation of the first-mentioned carbohydrate. Only when this substrate is all used up and the stationary or decreasing phase is reached can the second carbohydrate be attacked. The amount of growth at the end of the first cycle is proportional to the amount of the first carbohydrate. This type of curve is due to the induced formation of the carbohydrate-attacking enzymes. The carbohydrates in the first group, whose combination in pairs yields normal curves, are evidently attacked by *constitutive* enzymes, so that the cell's enzyme content does not change during growth. The ones in the second group are those whose enzymes are not initially present and are formed *inductively* by presence of the substrate. This induced enzyme formation is *repressed* by presence of the first carbohydrate. Thus, fructose represses formation of "arabinozymase", and glucose that of sorbitol dehydrogenase. It is indeed an old observation that adaptation of yeast to galactose is prevented by glucose, and the attack of starch by Penicillium is prevented by sucrose (Monod, 1947). In some cases, the second carbohydrate inhibits the formation of enzyme for a third, and a *triauxic* curve results.

Uptake of Solutes

Although it seems clear from the above that growth is limited by the uptake of solutes, and perhaps also of water, our knowledge of the uptake process is not extensive. For phosphate, uptake is known to be linked to metabolism. Yeast cells take up little phosphate in the absence of nutrient, and the inhibition of metabolism by cyanide, dinitrophenol, or azide leads to inhibition of phosphate uptake (Mullins, 1942; Hotchkiss, 1944). On the other hand, when riboflavin is added to yeast, it increases phosphate uptake without increasing the oxidation of carbohydrate; what *is* increased, however, is the conversion of glucose to cell substance, i.e., oxidative assimilation (Nickerson and Mullins, 1948). Hence the uptake of phosphate is increased by increasing its conversion to synthetic products. Unlike yeast, the bulk of the phosphate in *M. pyogenes* is in the inorganic (acid-soluble) form. By using P^{32} the uptake and loss of phosphate can be followed independently, and it is clear that there is rapid exchange

(Mitchell and Moyle, 1953). The exchange mechanism is poisoned by mercury, silver, and other heavy metals (Mitchell, 1954). During respiration the uptake is not appreciably increased, but the outgo is abolished, so that the net accumulation increases. The uptake of iron by C. *diphtheriae* differs from this picture in that it seems quite independent of the functions of the iron. Both Fe^{++} and Fe^{+++} are taken up rapidly, even if the iron level is high enough to inhibit toxin production completely; the shape and length of the growth curve are quite unaffected (Edwards and Seamer, 1960). Potassium, however, resembles phosphate in that its uptake depends on sugar metabolism, while sodium seems to pass in and out freely and not to be accumulated.

With yeast the K taken up is strictly proportional to the sugar added, and it is excreted again as the sugar is used up (Pulver and Verzar, 1940). Once taken up, the intracellular K promotes the uptake of other solutes, especially glycine (Eddy and Indge, 1962). Since K plays a special role in fermentation, and since also iodoacetate, which inhibits fermentation, inhibits K uptake, one might conclude that the K is taken up only to the extent that it can function.

However, the analogy with higher plants (Laties, 1959), in which sugar oxidation leads to the uptake and accumulation of both cations and anions, makes one very cautious about so simple an explanation. Indeed, in plants the uptake of water itself is linked to oxidative metabolism (Thimann and Leopold, 1955). The uptake of sugar is also dependent, at least largely, on metabolism.

With yeast it has been known for a long time that sugar is taken up faster than it can be metabolised, and after suspension in glucose solution, yeast becomes rich in glycogen (Stier and Stannard, 1936; cf. Brooks, 1947). The uptake can hardly be linked to glycogen formation, however, since it is too rapid. The speed of uptake is remarkable because glucose is not lipoid-soluble and, therefore, on classical views of permeability it would be expected to enter slowly. The same is true of amino acids, which are taken up rapidly, as shown below. Part of the explanation is that the classical concepts, according to which lipoid solubility and molecular size determine rates of entry, are largely founded on entry into the vacuole of plant cells, which is much slower than entry into the cytoplasm.

The temperature range the uptake is more than double the combined fermentative and oxidative breakdown. Nevertheless, uptake is entirely dependent on consumption, since it remains proportional throughout the temperature range (cf. the two upper curves), and it is about 90 per cent stopped by M/100 iodoacetate which inhibits both respiration and fermentation almost completely. Cyanide, which inhibits respiration but increases fermentation, does not prevent uptake and can even increase it. On the other hand, while metabolism virtually stops at 0°C, some glucose uptake persists, probably due to passive diffusion. Amino acids can also be taken up both actively and passively, and the process is complicated by their incorporation into proteins.

The active process depends on a source of energy; glucose promotes the uptake of glycine by yeast one-hundredfold (Eddy and Indge, 1962), while azide, iodoacetate, dinitrophenol, and fluoride all inhibit uptake of glutamate by both *M. aureus* and *Sc. fecalis* (Gale, 1951). In the first stages, the uptake is partly reversible, i.e., the amino acid can be exchanged for another, both in *E. coli* (Cohen and Rickenberg, 1956) and in Candida (Cowie and McClure, 1959), but this happens only very slightly in yeast. However, if yeast that has taken up C^{14}-phenylalanine for 5 or 10 minutes is extracted with hot water, the amino acid is seen to be still free; if protein synthesis is depressed by ultraviolet irradiation, some of the phenylalanine can be displaced by other amino acids (Halvorsen and Cohen, 1958). Thus, the exchangeability varies with the organism and conditions. The uptake of phenylalanine by yeast is depressed by valine (and vice versa), although valine does not stop incorporation of phenylalanine into protein; it merely inhibits the uptake process proper. Tyrosine or phenylalanine similarly inhibits the uptake of tryptophan in *E. coli* (Scott and Happold, 1962). This interference by related amino acids doubtless explains certain specific effects on growth; those Neurospora mutants that require lysine are inhibited by arginine, which has no effect on growth of the wild type (Doermann, 1944), and similarly a mutant requiring leucine is inhibited, competitively, by 8-chloroleucine (Ryan, 1952). In *Cl. septicum*, a pantothenate-requiring strain is inhibited by taurine (analogue of part of the pantothenate molecule), while the wild type is not (Ryan *et al.*, 1947).

Such specificity indicates that amino acids are taken up by rather specific carrier mechanisms, perhaps like those for galactosides. In the special case in which glutamate uptake into *M. aureus* is inhibited by cysteine or alanine, the glutamate is combined with the cysteine or alanine into a peptide, which is excreted (Gale and van Halteren, 1951); this suggests that both are "activated" by the carrier mechanism and in this state combine *with one another* and therefore, lose their hold on the carrier. At a later stage, protein is formed.

Even cells disrupted by supersonic vibration can still form proteins, although, having little "storage pool," they require other amino acids to be supplied as well (Gale and Folkes, 1955). A peculiar feature is that the most recently absorbed amino acids are converted to protein preferentially; when yeast is supplied with C^{14}-phenylalanine first, and then with C^{12}-phenylalanine, the radioactivity of the protein formed is lowered, although that of the free amino acid in the cell is sustained (Halvorsen and Cohen, 1958). The pool of absorbed amino acid, therefore, may not be on the direct pathway of protein synthesis.

Permeation Enzymes

Metabolism somehow enables the cell to bring certain organic compounds inside, must be set against the fact that some yeasts are unable to ferment specific sugars, although they contain the enzymes to do so. The autolyzates of many yeast strains contain maltase, invertase, etc.; yet the intact cells do not metabolise maltose or sucrose. Even stranger, a strain of *E. coli* could completely metabolise maltose but not glucose, although it contained hexokinase (Doudoroff, 1951). The only reasonable explanation for such phenomena is that there are highly specific permeation systems or carriers in the cell membrane, which, when activated by oxidative metabolism, take their particular substrates into the cell. In one case, that of the galactosides, this has been confirmed by Monod and his co-workers, and the system is called a "galactoside-permease." The uptake phenomenon can be shown in *E. coli* by first growing the cells in presence of lactose or other galactoside, so that they have formed the galactosidase enzyme adaptively.

If to these induced cells is added a thiogalactoside (i.e., with S instead of O in the galactoside bridge), this compound is at once accumulated to perhaps 100 times the external concentration, as can be shown by using the compound labelled with S^{35}. If the cells are then extracted, the S^{35} is found to be mainly present as unchanged thiogalactoside, and furthermore if an unlabelled thiogalactoside is added, some of the S^{35} comes out again (Rickenberg *et al.*, 1956). Now these thiogalactosides are not hydrolysed by galactosidase, and they are hardly likely to be bound in the cell, because the high internal concentrations reached (up to 5 per cent of the cell dry weight) would require an impossibly high content of the binding material.

If neither bound nor metabolised, then the thiogalactoside must be free in the cell, and hence the uptake system must act catalytically. The same would be true for other substrates also. That the substrate is truly free and contributes to the cell's osmotic pressure was ingeniously shown by using wall-less protoplasts, i.e., cells that had been treated with lysozyme. Protoplasts were prepared from cells that had been induced to form permease by being grown in lactose. When these were suspended in dilute lactose solution, they burst, showing that their osmotic pressure had been increased by the further accumulation of lactose (Sistrom, 1958). Suspension in glucose of the same molarity did not cause bursting. Protoplasts from uninduced cells also did not burst. As with the amino acids, azide or dinitrophenol inhibits the uptake, although they do not inhibit the actual hydrolysis of galactosides. Addition of the thiogalactoside even causes a very small *increase* in the endogenous respiration (Kepes, 1957). Since the thiogalactosides are not hydrolysed by galactosidase, the energy requirement must be to carry out the work of accumulating the compound against the concentration gradient. The evidence that the "permease" is enzyme-like (Cohen and Monod, 1957) is:

1. Its formation, like that of the galactosidase, and indeed of many other enzymes, can be induced.
2. Such induction has all the properties of a protein synthesis specifically here it is inhibited by chloramphenicol, by amino acid deficiency, and by certain amino acid analogues.
3. Like many other enzymes, it depends on an SH group for its activity and is inhibited by mercury compounds and other SH reagents.
4. Formation of the permease, like that of many other induced enzymes, is strongly inhibited by glucose. It is evident that formation of a specific permease can explain many quirks of metabolism such as the inability of intact cells to metabolise specific compounds for which they contain the enzymes.

This "cryptic" behaviour can indeed be imitated in *E. coli*, because several "cryptic" mutants have been found that are able to form galactosidase, as shown by extracting the cells, but nevertheless metabolise galactosides only extremely slowly (due to passive entry) (Herzenberg, 1959). A second type of mutant, which does not form galactosidase but can still take up galactosides, obviously produces the permease but not the hydrolytic enzyme. However, so far the evidence for permeases acting on other substrates is not extensive. They may participate in the uptake of some glucosides and glucuronides, and in one such instance, the accumulation of thiophenyl glucuronide, which is not metabolised, the process is inhibited by several closely related compounds, as well as by azide and dinitrophenol. Also the accumulation can go on simultaneously with that of galactosides, or separately, so that two independent permeases must be present (Stoeber, 1957). Nevertheless, it must be remembered that permeases are not absolutely essential since, particularly at high concentrations, passive diffusion allows appreciable entry of galaclosides (Holmes *et al.*, 1961). Indeed, in the case of galactosides, an enzyme for their hydrolysis is induced when the cells are exposed to melibiose or other galactosides, just as for galactosides; yet there was no evidence for the formation of a permease (Sheinin and Crocker, 1961). A group of permeases for organic acids appear to act in *Ps. aeruginosa*, for while these cells oxidised many acids only after a short lag period, their cell-free extracts showed no such lag in oxidation (Clarke and Meadow, 1959).

The lag is thus interpreted as being the time required for the synthesis of the specific permease. Many of the observations with amino acids above may also be due to specific permeases, but they seem not to be inducible, and the situation may be more complex. In animal cells, the postulated "carriers" for taking up solutes are generally thought not to be themselves enzymes, but rather products of enzymes forming parts of the cell membrane. The future will have to decide how far the action of permeases can be generalised to other solutes. The Synthesis of the Major Polymers

Linkage between Synthesis and Breakdown

The growth of bacteria represents the synthesis of complex compounds from simpler materials. Here we shall be concerned with general mechanisms of synthesis, primarily of polymers and storage materials, and especially with the crucially important special case of the synthesis of enzymes, on which all other metabolic processes of course depend. Hydrolysis of any large molecule to smaller units is a reversible reaction, and if the small molecules are in high enough concentration, some synthesis occurs. For example, fats are hydrolysed by lipase to glycerol and fatty acid, but a concentrated mixture of glycerol and fatty acid, in presence of lipase, synthesises some fat, and an equilibrium mixture is reached. Similarly a very concentrated mixture of amino acids in presence of chymotrypsin will synthesise true proteins (Tauber, 1951).

Few biological syntheses, however, are carried out in this way, because generally they go almost to completion in dilute intracellular solutions. In order to do this, since synthesis produces compounds of higher energy, another process must be coupled with it to supply the energy. This "exergonic" (energy-yielding) reaction is usually a breakdown. One of the simplest examples, in which the bond broken is of the same type as the one formed, and the amounts of energy required are extremely small, is the synthesis of polysaccharides from disaccharides.

A chemical survey of the dextrans formed by 96 strains of Leuconostoc, Streptococcus, Acetobacter, etc., revealed that they all have 1, 6 linkages with up to 40 to 50 per cent of 1, 4 and 1, 3 linkages as well (Jeanes *et al.*, 1954). All these are group-transfer reactions in which the transferred group is an entire glycoside, and the enzymes *transglycosidases.* The energy required for the linkage of one glycoside unit to another does not differ much whether there are 2 or 200 units in the molecule, so that the dissociation of the link in the small molecule furnishes enough energy to remake it in the large one. Formation of polysaccharide from monosaccharide is a quite different process. The synthesis of polysaccharides is of special importance in the formation of capsules, particularly on account of their interrelation with pathogenic virulence. However, it must be noted that capsular materials are seldom pure polysaccharides.

Phosphorylative Synthesis

Simple Cases: In more complex syntheses, the energy is produced by one reaction and utilised by another, e.g., in the synthesis of starch. When phosphorylase is added to glucose-1-phosphate (K salt), in presence of a little starch, starch is synthesised (Hanes, 1940). The synthesis proceeds in vitro, at room temperature and neutral *pH*. There is little energy change in this reaction, and indeed it is reversible. However, the normal synthesis of starch entails additional steps whereby the glucose-phosphate first combines with ATP to produce adenosine diphosphoglucose, ADPG, eliminating pyrophosphate, and it is the ADPG which then, in presence of the enzyme starch synthetase.

Instead of ADPG, uridine diphosphoglucose also acts, but at only one-tenth the rate (Leloir *et al.*, 1961). To form glucose-1-phosphate from glucose requires ATP, and to transfer it to starch requires a second molecule of ATP. If phosphorylase is used, only one ATP is required. In either case a continuous synthesis of polysaccharide can only occur if oxidation (or fermentation, or photosynthesis) continuously forms ATP. Thus, the overall reaction comprises an oxidation *linked* to a synthesis by ATP transfer. Synthesis of the disaccharide sucrose follows a similar pair of pathways: in *Ps. saccharophila,* sucrose phosphorylase combines glucose1-phosphate arid fructose into sucrose. As with starch the overall reaction requires an oxidation (as source of ATP, or ATP plus UTP) linked to the synthesis. The pattern is not limited to carbohydrates. Amino acids can enter into a variety of reactions by being first converted to carboxyl phosphates, which are very reactive. For instance, glutamine is synthesised from glutamic acid and ammonia by enzymes from liver or disintegrating cells of *M. aureus,* but only in the presence of ATP (Lipmann, 1949). The first step is formation of y-glutamyl-phosphate, which reacts with NH_3 or NH_2OH, liberating the H_3PO_4. In the synthesis of thiamine from its pyrimidine and thiazole precursors, *both* have first to be phosphorylated, the pyrimidine forming a pyrophosphate (Brown, 1960).

Protein Synthesis

The synthesis of proteins involves a much higher degree

of complexity, since many different amino acids must be linked in a particular order. The first stage is like that for pantothenic acid, namely, formation of the adenylate derivative, which is most probably bound to the enzyme (Hoagland *et al.*, 1956): The enzymes forming these aminoacyl-adenylates are a related group, each of which is specific for one amino acid only. They can often react instead with an amino acid analogue, however, such as ethionine or parafluoro-phenylalanine, and these become incorporated into the final protein; *Bac. subtilis*, grown in presence of ethionine, produces a crystalline amylase which contains ethionine instead of methionine (Yoshida and Yamasaki, 1959).

The second step is transfer of the aminoacyl residues to soluble RNA. This transfer does not require a second series of enzymes, but the enzyme-bound aminoacyl-adenylate apparently reacts directly with the RNA (Berg, 1958; Wong *et al.*, 1959). In bacteria, this occurs very rapidly, since the binding of C^{14}-amino acids to the RNA of growing *E. coli* reaches a maximum in about five minutes (Lacks and Gros, 1959). The reaction takes place at an adenylic acid residue which is attached to the rest of the RNA by two cytidylic acid residues; *i.e.*, at -AMP-CMP-CMP-RNA (Hecht *et al.*, 1958-1959; Zachau *et al.*, 1958). The linkage involves the 2' or 3' hydroxyl group of the ribose in the AMP, and if the RNA is hydrolysed by ribonuclease, the amino acid is recovered as its adenosyl derivative (Preiss *et al.*, 1959).

Since the whole process thus takes place at one end of the RNA molecule, the specificity for particular amino acids must reside in the rest of this large molecule. In the RNA of *E. coli*, there is only one terminal AMP residue for about 100 nucleotides, so that the whole subunit, of a molecular weight not less than 35,000, is used in attaching to one amino acid (Cohen and Gros, 1960; Hoagland, 1960). It follows, therefore, that the rate of turnover must be very high, the soluble RNA cycling rapidly. Since each different amino acid must attach to a specific RNA, it follows too that an apparently pure preparation of soluble RNA really contains a large group of slightly different molecules. Work on separating these specific "amino acid acceptors" is going on actively.

Evidently the reaction goes best if most of the 20 amino acids required to form the protein are transferred together to the particles (Hoagland, 1960). The transfer requires a single soluble enzyme, which is not specific for amino acids (unlike the RNA's), although it is specific for the organism, that from *E. coli* not functioning with the ribosomes from liver (Nathans and Lipmann, 1961). The soluble RNA, by contrast, is not specific for the organism at all; the mixture from *E. coli*, charged with amino acids, can react with the reticulocyte ribosomes of the rabbit to synthesise their normal product, hemoglobin. Besides the enzyme, guanosine-triphosphate, GTP, Mg+% and a phosphorylating system are needed (Nathans *et al.* 1962).

Thus, in protein synthesis ATP is needed for the very first step (formation of adenosyl-amino acid), and in addition the synthesis of each individual amino acid generally involves several phosphorylations. GTP also participates in the synthesis in some way. Protein synthesis is thus highly dependent on energy supply.

The *type* of protein formed is determined, not by the ribosome protein, but by another RNA, the "messenger-RNA" (abbreviated m-RNA), an unstable form whose composition is determined by the DNA, which acts as primer for its synthesis. What the messenger-RNA does is to arrange the aminoacyl-RNA's in a specified order. This order corresponds to its own composition, and it is now probable that three particular nucleotides compose a "code" which corresponds to a given amino acid. This fundamental conclusion was derived as follows: in the cell- free *E. coli* extract DNA-ase inhibits protein synthesis, because the m-RNA is then not formed (Tissieres *et al.*, 1960). Only if messenger- RNA is added can protein synthesis now be reinstated, so that we have here an *assay* for m-RNA. Using pure polyuridylic acid ("poly-U") formed by the Azotobacter enzyme, Nirenberg and Matthaei (1961) added various OMabeled amino acids in presence of ribosomes, ATP,

GTP, Mg, substrate, and the enzyme. With phenylalanine, and with *no other* amino acid, the resulting protein was radioactive. About 1.5 moles of uridylic acid in a "poly-U" of molecular weight about 50,000 incorporated 1 mole of phenylalanine into protein. On the assumption that three nucleotides are needed, the code would be UUU.

Polyadenylic acid, "poly-A," alone in similar experiments formed no protein, but polymers containing two or three different nucleotides each caused incorporation of only one or two amino acids. Thus, poly-UA incorporates lysine and acts best when the A content is high; therefore, the code for lysine is probably UAA. Present data lead to the following tentative nucleotide combinations as codes for the amino acids (Matthaei *et al.*, 1962): Some of these overlap; some are uncertain. It has been objected that the proportion of uridylic acid in a nucleic acid that would synthesise normal protein would be impossibly high (Chargaff, 1962). However, some of the U's may be only acting as "spacers" to maintain the other two members of the triplet in the right positions (Jukes, 1963), and indeed Ochoa (1963) finds that poly-A can to some degree promote the incorporation of lysine, and poly-C of proline. Thus a firm basis for the coding system is clearly in sight.

One important point, namely, that the code selects aminoacyl-RNA's and not free amino acids, has been rigidly confirmed. Since cysteine can be chemically hydrogenated to alanine, C^{14}-cysteyl-s-RNA was carefully hydrogenated to C^{14}-alanyl-s-RNA. This was added to a preparation of *E. coli* ribosomes in presence of poly-UG, which incorporates cysteine but not alanine. The resulting protein contained C^{14}, i.e., it had incorporated alanine; this means that because the alanine was attached to the RNA specific for cysteine, it was "mistaken" for cysteine by the coding system. Oxidation of cysteyl-s-RNA to the cysteic acid derivative similarly caused incorporation of cysteic acid into protein (Chapeville *et al.*, 1962). Another valuable confirmation comes from the nitrous acid experiments on virus. For cytosine would be converted by HNO_2 to uracil, so that UUC would be replaced by UUU, and CUC by UUC.

Correspondingly one mutant formed by HNO_2 has its serine replaced by phenylalanine, and another has proline replaced by leucine (Martin *et al.*, 1962). Similar changes have been detected in haemoglobins formed by cell-free systems (Smith, 1962). The ribosomes are not the only site of protein synthesis, because fractions rich in phospholipid also incorporate C^{14}-amino acids (Scha-chtschabel and Zillig, 1959, with *E. coli;* Godson *et al.*, 1961, Hill, 1962. with *Bac. megateriurri).* These fractions derive from the celi membrane, since they can be prepared from washed isolated membranes, and in the Bacillus, they become labelled 10 to 30 times as intensely as the ribosomes in the first few minutes. Thus, the soluble RNA may orient itself along more than one kind of receptor surface.

Synthesis of Nucleic Acids

This is a subject of very recent rapid growth in understanding. It stems from the finding of Ochoa and co-workers (1957) that cells of *Azotobacter vinelandii* yield an enzyme able to polymerise nucleotidetriphosphates to RNA-like macromolecules. The purified enzyme had the same activity for all the nucleotides tested and is therefore, a single enzyme; nevertheless, it converts ATP to polyadenylie acid, GTP to polyguanylic acid, UTP to polyuridylic acid, and CTP to polycytidylic acid. The reaction lags at first, and the lag is removed by adding a small amount of polymer as primer; each polymer primes only for the formation of itself, except polycytidylic acid which acts as primer for all four. Mixed polymers can also be made. *M. lysodeikticus* gives a similar preparation (Beers, 1957), which is easier to extract (Boldingh and Veldstra, 1961). These preparations have played a crucial part in the experiments on "coding" mentioned above. The composition of the RNA made by the *E. coli* enzyme can be influenced by DNA used as primer; if polymers of a single deoxyribonucleotide are used as primers, they cause synthesis of the corresponding ribonucleotide polymer, deoxyribothymidylate resulting in riboadenylate, etc. (Chamberlin and Berg, 1962). *Bac. megaterium* yields an even more striking preparation, since, when its protoplasts are lysed with detergent, a nuclear fraction, containing almost all its DNA, is obtained and this directly synthesises RNA from a mixture of the four triphosphates (Godson and Butler, 1962).

In reactions like this, one can see just how the nucleus controls the biochemistry of the cell. A parallel system obtained from *E. coli* synthesises DNA from a mixture of the four deoxynucleotide triphosphates (Lehman *et al.*, 1958, Aposhian *et al.*, 1962). Again a primer of a little DNA is needed; otherwise the reaction shows a marked lag. The DNA acts as a nucleus for autocatalytic replication (Radding and Kornberg, 1962). This system differs from that for RNA in that it needs all four nucleotides (thymine, cytosine, adenine, and guanvie deoxyriboside triphosphates), and if one is omitted, the polymerisation becomes exceedingly slow. However, it can, in the absence of a primer, form a co-polymer of adenine and thymine deoxyribc sides only. This product then acts as primer in formation of itself, and its properties are like those of a modified DNA (Schachman *et al.* 1960).

The above brief outline of our explosively growing knowledge of the synthesis of the proteins and nucleic acids will serve as background for the treatment of the formation of assimilate and of enzymes, and the transfer of genetic material, which follows.

Oxidative Assimilation

The overall synthesis of polymers in intact bacteria of course comprises many of the individual processes. That it depends on oxidation for energy source was first brought out by Cook and Stephenson (1928) with *E. coli.* When various oxidisable substrates were added to the washed cells, the increased oxygen consumption ceased after a time, although only about two-thirds of the calculated amount of oxygen had been consumed. The enzymes appeared to be still intact, and no intermediate, unattackable by the bacteria, had seemed to accumulate in the medium. Evidently the rest of the substrate had been converted to polymer, or "assimilated." Cells whose power of multiplication had been largely reduced (by exposure to ultraviolet) behaved in nearly the same way. Spirilla accumulate granules of "volutin," an assimilation product, when the cells (in the absence of nitrogen compounds and thus unable to grow) oxidise various substrates; Giesberger (1936) found that the oxygen taken up was never enough for complete oxidation of the substrate, so that part was oxidised and part "assimilated." Assuming the assimilate to be of the composition of a carbohydrate (CH_2O), the oxygen consumption indicated the following overall reactions:

$$\text{Acetic } CH_3COOH + O_2 \longrightarrow (CH_2O) + CO_2 + H_2O$$

$$\text{Propionic } 2\ C_2H_5COOH + 5\ O_2 \longrightarrow 2(CH_2O) + 4\ CO_2 + 4\ H_2O$$

$$\text{Lactic } CH_3CHOHCOOH + O_2 \longrightarrow 2(CH_2O) + CO_2 + H_2O$$

$$\text{Pyruvic } CH_3COCOOH + \frac{3}{2}\ O_2 \longrightarrow (CH_2O) + 2\ CO_2 + H_2O$$

Similar results with the colourless alga *Prototheca zopfii* (Barker, 1936) show that organic acids, as well as glycerol and ethanol, are assimilated by the cells, the product having the composition of a carbohydrate. The gas exchange measurements agree acceptably with the proposed equations, particularly since the assimilate may not be exactly of the composition (CH_2O). They also agree well with the assimilation processes of the spirilla. These data suggest that all the reactions proceed via acetate (Anderson, 1945). Acetate is no doubt utilised via the glyoxylic acid cycle. Other organisms assimilate with not quite the same relationships. For instance, *Ps. calcoacetica* and *E. coh* respiring acetate, use 14 moles of oxygen (Cook and Stevenson, 1928; Clifton, 1946):

$$2\ CH_3COOH + 3\ O_2 \longrightarrow (CH_2O) + 3\ CO_2 + 3\ H_2O$$

while the water mould *Leptomitus lacteus* uses only *per cent* mole of oxygen (Schade and Thimann, 1940):

$$3\ CH_3COOH + 2\ O_2 \longrightarrow 4(CH_2O) + 2\ CO_2 + 2\ H_2O$$

On glycerol, Mycobacteria use 1 to 1.5 moles of oxygen, and there is good evidence that the glycerol is converted to pyruvate (Hunter, 1953; Clifton, 1946). Carbohydrate is also in part assimilated. With yeasts oxidising glucose, only about one-third the theoretical amount of oxygen is taken up, the reaction being:

$$C_6H_{12}O_6 + 2\ O_2 \longrightarrow 4(CH_2O) + 2\ CO_2 + 2\ H_2O$$

Not all the (CH_2O) could be identified as actual carbohydrate, about half of it, in *Sacch. cereviseae,* being in the form of other compounds, probably fat (Stier and Newton, 1939). Endomyces and Geotrichum in particular build up large amounts of fat. The assimilation product may vary with the substrate, some substrates giving rise predominantly to fat, others to protein. The amount of fat and sterol formed depends on the content of CoA and can be varied in yeast by varying the pantothenic acid in the medium (Klein and Lipmann, 1953). It can also be varied in algae by changes in growth conditions. Much of the assimilate, in bacteria at least, is evidently a polymer of -hydroxybutyric acid. This compound is also accumulated by photosynthetic bacteria. When *Ps. saccharophila* assimilates C^{14}-labelled glucose, two-thirds of the C^{14} is found in this polymer (Doudoroff and Stanier, 1959). Assimilation on acetate and butyrate gave over 80 per cent conversion of their C^{14} to poly-hydroxybutyric. Oxidative assimilation in presence of a nitrogen supply leads to protein synthesis. In the water mould *Leptomitus lacteus,* ammonia is absorbed from the medium when acetate or butyrate is being assimilated, and the product has about the overall composition of alanine, $C_3H_7NO_2$ (Schade and Thimann, 1950).

If the product is really protein, it would mean that leucine (or acetate and ammonia) gives rise to all the amino acids. *Ps. aeruginosa* similarly takes up NH_3, while organic acids are being oxidised (Bernheim and de Turk, 1951).

In general, the rate at which an organism can grow is doubtless a function of the rate at which it can assimilate, and especially of the rate at which it produces enzymes. Slow-growing mutants of Salmonella, especially *S. pullorum,* were found to oxidise alanine, serine, and glycine completely, without any assimilation (Stokes and Bayne, 1958, 1961); while on some other amino acids, oxygen uptake stopped when 70 to 80 per cent of the theoretical O_2 had been taken up, indicating slow assimilation. The normal fast-growing strains all stopped oxidising alanine and serine when 50 to 60 per cent of the theoretical O_2 had been taken up, which about corresponds to the types of assimilation above.

Assimilation is readily prevented by poisons without poisoning the remainder of the respiration. Fluoride prevents yeast from converting glucose to glycogen, probably by poisoning the phosphoglucomutase that converts glucose-6-phosphate into glucose-1-phosphate (Nickerson and Chung, 1952), Iodoacetate, azide, and especially dinitrophenol, all prevent the assimilation of acetate by *Ps. calcoacetica* (Clifton, 1946); in their presence, therefore, the oxidation of acetate goes almost to completion.

Since the nitrophenols are specifically known to inhibit phosphorylation, this confirms the need for ATP in these processes. In *Mycob. ranae,* iodoacetate prevents the oxidative assimilation of pyruvate, which instead is oxidised to acetate and CO_2 (Lindsay *et al.*, 1950). On the other hand, arsenite prevents the oxidation of the pyruvate so that in its presence Mycobacteria convert pyruvate to glycerol (Hunter, 1953). As would be expected, all these assimilation inhibitors inhibit growth also.

Only in those cells which, because of metabolic imperfections or presence of poisons, are neither growing nor assimilating, can true respiration uncomplicated by synthesis be said to go on. In all other cases, respiration is only a partial aspect of the vital process; as Kluyver once put it, the CO_2 evolved resembles the chips scattered when the sculptor converts a shapeless block of granite to a desired form.

Fermentative Assimilation

Assimilation can also, of course, occur anaerobically when microorganisms grow without oxygen. In yeast that is fermenting sugars (in the absence of oxygen), only about 70 per cent of the sugar is accounted for as CO_2 and alcohol, most of the rest being found as increased weight of cells (Meyerhof and Schulz, 1936; van Niel and Anderson, 1941).

Young yeast cells, even in air, ferment sugars and consume very little oxygen (Swanson and Clifton, 1948). As the sugar becomes used up, the assimilate, as well as the alcohol produced, begins to be oxidised, and the resulting energy of oxidation leads to further assimilation. Similarly *Propionibact pentosaceum,* while fermenting dextrose, stores 10 per cent of the carbon as polysaccharides and ferments this again when the dextrose is all used (Wood and Werkman, 1934). Several bacteria store glycogen, both aerobically and anaerobically, and use it up again if starved.

On the other hand, *Sc. fecalis* shows no fermentative assimilation at all and gives an almost theoretical yield of lactic acid from glucose (van Niel and Anderson, 1941). This may be connected with the great need of the lactic bacteria for preformed amino acids and vitamins. The high yields of lactic acid often recorded in the literature doubtless rest on similar absence of assimilation.

Formation of Enzymes

Basic Observations: When a medium containing a new substrate is inoculated, the growth of any mutant cells that can attack the new substrate is immediately favoured. Similarly, if a required nutrient is missing, the growth of mutant cells that can synthesise this nutrient is favoured. In this way, a bacterial population can be continuously variable, since three factors favouring rapid change are present: high selection pressure exerted through essential nutrients, a high mutation rate, and enormous populations. However, in addition to selection, there is a quite other factor operating, namely, the formation of specific enzymes *in response to the substrate* and independent of growth. The study of this phenomenon, called adaptation or induction, has shed a flood of light on the process of enzyme formation in bacteria. The first clear instance of this process came from the work of Stephenson and Stickland (1933) with formic hydrogenlyase.

This enzyme-complex appears in *E. coli* when it is grown on a medium containing formate. If cells grown without formate are washed and suspended in formate, the enzyme is produced within two hours; the only requirement is the presence of broth, since adaptation occurs poorly in formate alone. However, no growth need occur. In experiments lasting three hours, the amount of enzyme, measured as the rate of H_2 evolution per milligram bacterial N per hour (= Q_{Hj}) rose from 0 to about 1000, while the number of cells (*ca.* 3 X 10^8 per millilitre) showed no increase at all.

The formation of enzyme thus took place entirely in the lag phase. Instead of broth, the amino acids, serine, aspartic acid, and arginine, will suffice to supply whatever energy or nitrogen is needed (Pinsky and Stokes, 1952). Induction can even occur anaerobically, providing fumarate or nitrate is added, evidently to act as hydrogen acceptors for an energy-yielding reaction (Kushner and Quastel, 1953). Tryptophanase, the enzyme that converts tryptophan to indole, is similarly formed when *E. coli*, grown without tryptophan, is suspended in this amino acid. No other substrate is required, and no growth occurs, although tryptophan itself is, of course, a potential energy source (Happold *et al.*, 1936, 1941). In yeasts the fermentation of galactose can be similarly induced; energy, if needed, is supplied by endogenous substrates, and on pure galactose alone the galactokinase that forms galactose-1-phosphate and is the limiting factor in galactose utilisation appears in about an hour (Lindegren and Pelleroni, 1952; Sheffner and McClary, 1954). Another remarkable induction occurring in yeasts is the formation of cytochromes (Ephrussi and Slonimski, 1950). After growing anaerobically for 24 hours (six generations), yeast loses the ability to oxidise glucose, and the spectral bands of cytochromes *a*, *b*, and *c* disappear, *a* and *b*± replacing them. Aeration in glucose-buffer brings back both the full cytochrome spectra and the respiration. The change begins immediately, although, if the cells are in the logarithmic phase, there is an hour's lag, probably due to the synthesis of the oxidase (Lindenmayer and Estabrook, 1958).

In the case of nitratase, i.e., the enzyme in *E. coli* needed for the reduction of nitrate by organic hydrogen donors, enzyme formation occurs in nitrate with any metabolisable substrate, such as formate or lactate, but is fastest with sugar, probably because it is the best energy source (Pollock, 1946). In presence of glucose, the rate is further accelerated by adding amino acids and depends less on their *concentration* than on the number of different amino acids, which clearly indicates the need for protein synthesis. A similar case is formation of S-galactosidase by Aerobacter (induced by adding lactose); this proceeds for a few hours in buffer alone, but by 20 hours, the enzyme can no longer be formed.

If peptone is added with the lactose, the enzyme is formed at the full rate even after 68 hours' starvation, so that it is clearly amino acids that are needed (Strange, 1961). Such starved cells lose both protein and RNA to the medium. In the above cases, formation of the enzyme is induced by its substrate: formate, nitrate, tryptophan, galactose, lactose, or, in the case of the cytochromes, oxygen. However, in *E. coli* the galactosidase can be induced not only by lactose but also by the unattackable compounds melibiose (an a-galactoside), several thio-galactosides, and even weakly by galactose. Conversely phenyl/ J-galactoside, which *is* hydrolysed, does not induce formation of the enzyme (Cohn, 1957). In *Proteus vulgaris*, formation of leucine decarboxylase is induced by alanine, which is not attacked, almost as rapidly as by valine or leucine, which are substrates for the enzyme (Haughton and King, 1961). Also, some *E. coli* mutants are simultaneously induced to form a second protein (PZ), which is enzymatically inactive and was only detected by immunologic methods. In many cases, the inducer also induces formation of the "permease" described above, which may be a third protein. Thus, (1) the process of induction can be multiple, and (2) it has a different specificity from that relating the enzyme to its substrate. Formation of an induced enzyme is actual *synthesis* of the enzyme protein and not merely unmasking or activation of some inactive precursor. This has been shown in several cases, but most clearly as follows (Rotman and Spiegelman, 1954; Hogness *et al.*, 1955): *E. coli* was grown in presence of C^{14}-lactate or S^{35} sulphate, so that the cell proteins were all labelled to varying extents. The cells were then washed and transferred to isotope-free medium, to which an inducer for galactosidase was added.

After an interval, the galactosidase formed was extracted and purified; in both cases it was virtually free from radioactivity. Hence it must have been formed anew from the *unlabelled* constituents of the second medium. In growing cells, the enzymes once formed are not appreciably broken down again. During the lag phase, or in non-growing cultures, however, there is appreciable breakdown with resynthesis of protein, amounting to about 5 per cent of the cell proteins per hour (Mandelstam, 1958). Essentially the same is true for nucleic acids, both RNA and DNA being very stable in growing cells; while in starved *E. coli,* the rate of turnover of RNA is around twice as fast as that of protein. Such slow turnover accounts for the occasional formation of induced enzymes by nitrogen-starved cells (Wainwright and Nevill, 1956). The power to form an enzyme in response to an inducer is inherited as a normal genetic character. The earliest genetic work, e.g., on 3-galactosidase of *E. coli,* distinguished only ability and inability to use lactose.

Inhibition and Repression

The diauxic curves indicates the inhibition of the arabinose-attacking enzyme by glucose, etc. Clearer evidence that enzyme synthesis can be specifically inhibited was given first for tryptophan-synthetase, which combines indole and serine to form tryptophan (Monod and Cohen-Bazire, 1953); formation of this enzyme is inhibited by tryptophan and some of its analogues. Many such cases have now been studied. Sometimes, it is the direct products of the reaction that inhibit, but in others, it is the *end products* of the resulting entire sequence of enzymatic reactions. As examples, *valine* and *isoleucine* are formed by transamination between their keto-derivatives and phenyl-alanine or alanine; the keto acids, in turn are derived from pyruvate and a-ketobutyrate, respectively. Now a-ketobutyrate is formed by deamination of threonine, and this deamination is inhibited competitively by isoleucine (Umbarger and Brown, 1958). Thus the end product, isoleucine, powerfully inhibits the very first reaction in the sequence, a process called "feedback inhibition":

[1] semi-aide- [2] 5-carbox-hydeylic acid

Again, *threonine* comes from aspartate via its semi-aldehyde, and this is formed by the phosphorylation of one carboxyl. The enzyme for this, £-aspartokinase, is competitively inhibited by threonine (Wormser and Par dee, 1958). In these three cases, the control is thus by inhibition of the *enzyme activity*. In several cases, the control system is more complex, acting by preventing *formation* of the enzyme.

In *Zymob. oroticum* two of the enzymes are inducible; in *E. coli* strain 6386, which requires pyrimidines for growth, all three are very weakly present, but when the cells are washed and aerated *without* uracil, No. 1 increases 150 to 500 times, No. 2, 10 to 40 times, and No. 3, 4 to 10 times. At the same time, the total protein synthesis stops. Now chloramphenicol also stops total protein synthesis, but since it does not *cause* formation of these enzymes, their formation cannot be the result of stopping protein synthesis but must be the result of *absence of uracil*. If the cells are suspended in C^{14} leucine and then the proteins separated, high C^{14} activity is found in enzyme No. 1, showing that it has been actively synthesised. Hence uracil (or its products) inhibits formation of these enzymes. Another well-studied case is that of acetylornithinase (A-O-ase) which converts acetylornithine to ornithine and thus, in *E. coli*, lies on the path of synthesis of arginine. Mutant arginine-requiring or acetylornithine-requiring cells, when grown on arginine and transferred to an acetylornithine medium, grow only after a lag of some hours (Vogel, 1957). Such behaviour might appear to be due to induction of the A-O-ase by its substrate, but no evidence for such induction could be found. Instead, presence of arginine in the cells (also in wild-type cells) *lowered* the amount of the A-O-ase. It was deduced that arginine, end product of the sequence, *represses* formation of the A-O-ase which acts at an earlier stage.

Such repression has obvious value in economising on protein, and Vogel concluded that "a cell in an environment that supplies arginine will utilise this exogenous arginine, and conserve its resources by sharply curtailing the *now unnecessary* synthesis of A-O-ase." Looking at it the other way round, the formation of an enzyme that helps synthesise a necessary food, when that food is scarce or absent, makes growth possible on otherwise inadequate media. Thus induction and repression are two comparable ways in which synthesis of enzymes is controlled by their products. In at least two cases, the histidine and tryptophan-forming sequences, the control extends to a whole group of enzymes.

In *Salmonella typhimurium,* six enzymes function in the histidine series, and the activities of at least the last four bear a constant ratio to one another, irrespective of the type of mutant. Some 40 mutations in different locations give rise to cells with the same ratios of these AAAP enzymes. Furthermore, it is possible to grow a histidine-requiring mutant without adding histidine, by the device of supplying formyl- histidine, which is hydrolysed so slowly that (at any one time) very little free histidine is present in the cell. This minimises the repression, and it is then found that the concentrations of *oil four* of the enzymes studied increase by some 15 times (Ames and Garry, 1959; Ames *et al.,* 1960). Evidently, therefore, their formation has been repressed by histidine. What makes this particularly interesting is that genetic evidence, from recombination, shows that the genes controlling formation of the series of enzymes are located very close together on the chromosome, and indeed their order along the gene-string is almost *the same as the biochemical order* in which the enzymes have to function (Hartman *et al.*, 1960). This holds for five of the mutant groups, controlling, respectively, five of the enzymes and has been verified for over 200 individual histidine mutants of Salmonella.

A comparable sequence functions in the synthesis of tryptophan by *E. coli* and *S. typhimurium* (Smith and Yanofsky, 1960). The first enzyme forms anthranilic acid, the second links a phosphoribulose group to it, the third (probably dual) closes the ring to form indole-3- glycerol phosphate, and the fourth (doubtless multiple) reduces and aminates this to tryptophan. An alternative (fifth) enzyme splits off the side-chain to form indole, which combines with serine to form tryptophan also: Here, in *Salmonella, E. coli,* and *Bac. subtilis,* all the effective genes are closely linked and lie approximately in order of the biochemical sequence of their enzymes. An odd exception is the fact that in *Bac. subtilis* (where the gene order was determined by using transforming substances), the order of enzymes 2 and 3 is reversed (Anagnos- topoulos and Crawford, 1961). An instance from another field is the fermentation of galactose:

> Three enzymes are required to phosphorylate the galactose, combine it with UDPG, and reverse the spatial arrangement at carbon 4. Yet a single-step mutation, occurring both in *E. coli* K12 and in a Salmonella strain, can inhibit synthesis of all three (Kalckar *et al.*, 1959; Nikaido and Fukasawa, 1961).

Rhamnose offers a comparable example, with only two enzymes (Englesberg and Baron, 1959). Thus the "single mechanism," sometimes at least, controls a whole group of enzymes, which are closely related both in their substrates and in their spatial positions.

Finally, it is important that enzyme repression is common in synthetic *sequences.* It has now been found in nine bacterial syntheses, namely, of ornithine, citrulline, histidine, proline, methionine, valine, tryptophan, pyrimidines, and purines. A number of enzyme inductions are strongly affected by glucose. Formation of proteinases, amino acid oxidases, histidinase, and many other deaminating enzymes is suppressed by glucose, while formation of amino acid decarboxylases, tryptophan-synthetase, and several other synthetic enzymes is promoted by glucose.

The phenomenon of enzyme repression has been invoked to explain these glucose effects as follows (Neidhart and Magasanik, 1956). Enzymes whose formation is *repressed* by glucose would be those whose products can be readily formed by the cell by alternative routes directly from glucose, e.g., urocanic acid and the a-keto acids. Glucose would be metabolised so fast that these products would accumulate and repress the deaminases, etc., that produce them. Enzymes whose formation is *promoted* by glucose would be those whose products are removed by reacting with the glucose breakdown products (e.g., amines, or compounds which can enter a synthetic reaction if energy is supplied). In this case, the removal of the products would lift the repressive influence that these compounds normally exert. The theory remains unproven, though attractive.

Are Enzyme Induction and Repression Really Different?

For some time, the feeling has been growing that induction and repression are not two "comparable ways of controlling enzyme synthesis, but that they are really the same, i.e., that induction is *the lifting of a repression.* The best evidence for this is the following. The synthesis of S-galactosidase in *E. coli,* which is inducible in some strains, is constitutive in others, and the control resides in a pair of genes, i^+ and r, respectively. The conjugation phenomenon has been used to study inheritance here and has revealed two other factors besides the i gene; y, which makes a "permease.," or factor allowing galactosides to enter the cell, and z, which controls synthesis of the enzyme itself; z^- cells cannot make the enzyme under any conditions.

Thus: z^+ allows enzyme synthesis, z^- prevents it; y^+ allows permease synthesis, *y*⁻ prevents it; requires inducer for the formation of both galactosidase and permease ("inducible"); r allows synthesis of these in the absence of inducer ("constitutive"). The galactosidase and permease are in most cases made together, and in over 100 r mutants both enzyme and permease are formed (Jacob and Monod, 1961). All occurrences of z", y, and i^+ are found by crossing over to be located near together on the chromosome. Suggestively, in cells diploid for the i gene, when i^+ and r occur in the same cell, the cell is not constitutive for the enzyme, but requires inducer, i.e.. *inducibilitv is dominant.*

This means that the inducible is the active form, which does not fit with the idea that inducible cells are missing something needed to make the enzyme; instead, it suggests that they actively *produce* something that prevents enzyme formation. Only an inducing substance (on this concept) can stop this and hence allow the enzyme to be formed. Ingenious proof that this is the case, i.e., that the inducer removes a repression, has been given by timing the transfer of the male characters into the female during conjugation (Pardee *et al.*, 1959). Three characters were concerned, z, 1_9 and one controlling sensitivity or resistance to streptomycin, designated $5m^8$ and $Sm^T{}_9$ respectively. The mating was between: male, z+z+Sra^8, and female, *z~i~Sm*ʳ. The addition of the antibiotic streptomycin to the medium prevents the male cells from multiplying, but they can still conjugate. The female cells make no enzyme anyway. Thus, any increase in the enzyme in the culture must be due to *zygotes* that have received the z+ gene. The two cultures are mixed, and samples taken from time to time for £-galactosidase assay. In 20 minutes z^+ enters the females, as shown by formation of enzyme at once. Since no lactose is present, the enzyme is being made constitutively (r). But 60 minutes later. its formation stops; now it will resume only on addition of lactose (or other inducer); evidently i^+ has now entered. Hence the i^+ gene has *stopped* the formation of the enzyme which was proceeding normally; i^+_9 therefore, causes formation of a *represser.* Inducing substances prevent the repression.

Nature of the Represser

Since mutations of the gene do not affect other enzyme systems, it follows that the represser is highly specific. On the other hand, it can control more than one gene at a time; e.g., those for galactosidase jmd permease, in the above instance. In the case of histidine synthesis all the eight sequential enzymes concerned can be prevented from forming by a *single* mutation. Hence, the represser does not act upon a gene but upon something else which may control a group of genes located close together an "operator."

The above conjugation experiment, if methyl-tryptophan is added before *z+* enters, the synthesis of tryptophan is inhibited, and therefore, proteins, including -galactosidase, are not formed. However, represser is still formed, because if tryptophan is added 60 minutes after the methyl-tryptophan, to reinstate synthesis of the enzyme, the synthesis *will not start* without inducer. Thus, the represser is not a protein and therefore, not an enzyme. This rules out one possible explanation of repression, namely, that the represser is an enzyme destroying the inducer. The operator lies on the chromosome and controls the activity of the string of functionally related genes, which are close together. Each gene forms a specific ribonucleic acid molecule which migrates of protein synthesis and there forms a specific enzyme. That the step from gene to enzyme is mediated by a ribonucleic acid is to be expected, since RNA controls protein synthesis. Also cells which have been x-rayed sometimes fail to form enzyme, and the most effective wavelengths are those absorbed by nucleic acids (Pauly, 1959;). From the x-ray dosage needed to inactivate the synthesis, the calculated size of the "target," both for lysine decarboxylase and for galactosidase, gives a molecular weight of 20 millions, which compares well with that of the DNA of bacteriophage.

The operator may combine with a represser (probably a nucleic acid too) and thus be incapacitated. But an inducing substance, which similar to the enzyme substrate, can combine with the represser and prevent its attachment. As Vogel first suggested (1957): "Inducers or repressers have affinity for the very portions of the genome that control the formation of the corresponding enzymes." The evidence of Pardee *et al.* seems to make a mechanism of this type extremely general in bacteria.

Some Additional Points

1. Repression of formation of a related group of enzymes, al- though a striking phenomenon when it does occur, is not necessarily the rule. In some sequences, as in that for uracil above, formation of each enzyme appears to be repressible separately. An operator must in that case be assumed for each gene, which makes the above picture more complex.
2. The repression of enzyme *formation* must be clearly distinguished from the inhibition of enzyme *activity*. For instance, in the uracil-forming sequence described above, the first reaction is the linkage of carbamyl phosphate to aspartic acid; this reaction is inhibited by cytidylic acid (Pardee and Yates, 1956). Since cytidylic acid is formed from uracil, this means that the last product of a sequence inhibits the first step, another instance of the "feedback inhibition". Yet this does not concern enzyme synthesis, but only enzyme activity.
3. A controversial point concerns the relation between induced enzyme formation and growth. In *E. coli* the formation of galactosidase was first found to be strictly dependent on growth, and the enzyme represented a constant fraction of the total protein formed (Monod

et al. 1952). Later the enzyme was found to appear in a disproportionately large amount during the early phase of growth, so that, although protein was being synthesised, the enzyme was being formed preferentially. An external energy source, rather than an external nitrogen source, is necessary, i.e., the enzyme can be formed from reserve free amino acids. For example, *Ps. saccharophila* forms amylase well within an hour when suspended in buffer with starch, although S^{35}-sulphate was not incorporated detectably into protein in that time (Schiff *et al.*, 1959). There is, in fact, a small pool of free amino acids in *E. coli,* and this increases when the cells are nitrogen-starved, presumably by proteolysis (Mandelstam, 1958). A rate of protein "turnover" scarcely detectable by chemical means may suffice for formation of appreciable enzyme. Hence, neither growth nor net protein synthesis really needs to be a prerequisite for enzyme formation.

4. Correspondingly a very small amount of inducer can be effective. *Bac. cereus* makes penicillinase, the enzyme that hydrolyses penicillin, inductively, and by using S^{35}-labelled penicillin, it was shown that the isotope is fixed in the cells and that each molecule of penicillin fixed causes the formation of at least 10 molecules of enzyme (Pollock and Peret, 1951; Pollock and Torriani, 1953). However, if after a few minutes the cells are washed free of penicillin, the enzyme continues to be formed. This is no doubt because the cells fix and retain some penicillin (i.e., inducer) irreversibly. By contrast, with the galactosidase of *E. coli,* removal of the inducer stops enzyme formation in a few minutes (Cohn, 1957), and some other systems behave similarly.
5. So far, clear evidence for repression is limited to bacteria, although *induction* certainly occurs in yeast, Neurospora, and even in mammals.

6. Lastly, a striking recent development is the synthesis of £-galactosidase in an entirely cell-free system (Novelli *et al.*, 1961). This was discovered via studies of the effect of radiation on enzyme formation. The synthesis requires both particles and supernatant fraction from *E. coli* cells previously induced with thiomethyl-galactoside; amino acids, ATP, and several nucleotides are also needed. Within an hour the enzyme activity was more than doubled, and C'Meucine was generously incorporated. The reaction is inhibited by x-rays and ultraviolet, and then reinstated by adding DNA from untreated cells. Doubtless many other such systems will soon be elaborated.

3

Oxidative Process

Fermentations and Aromatic Metabolism

The asteur's definition of fermentation as life without air obviously cannot apply to any oxidative process. However, the oxidations with which we are here concerned involve many of the same substrates as the fermentations and have so much in common chemically with them that the contradiction in terms ("oxidative fermentation") is partially justified. In any event, the term *fermentation* is sometimes used now, especially in industry, to indicate any biochemical process carried out by microorganisms. Of the microbial oxidations that are related to fermentation, three types are important: (1) The conversion of ethanol to acetic acid, (2) The oxidation of hexoses and sugar alcohols to related acids and ketones, and (3) The conversion of sugars to citric, fumaric, and oxalic acids. These last are more characteristic of fungi than of bacteria and so will not be discussed here.

Acetic Bacteria

The flora of wine or beer exposed to the air and turning sour (acetifying) were studied by Pasteur in the 1860's in response to requests from the wine makers of France. In his "Etudes sur le Vinaigre" (1868) Pasteur describes two forms of "*Mycoderma aceti*" one forming a delicate dry veil on the surface of the liquid, the other producing "a sort of moist skin, swollen, gelatinous and slippery" and finally filling the whole liquid. Although Pasteur considered they were not bacteria, the name *Mycoderma* is unfortunate since it really applies to a group of non-sporeforming yeasts which are also found on the surface of souring wines.

These are aerobic, non-fermentative, budding cells, described by Persoon as early as 1822, and now classified in the *Torulopsoideae*. They oxidise sugars, alcohol, and organic acids. Other organisms, sometimes found in souring wines are true yeasts of the genus Zygosaccharomyces, and also, if the acetic acid content is not above 1 M, a group of nematodes, *Turbatrix aceti*. Little is known about the metabolism of these interesting animals. They can survive *pH.* values from 1.6 to 11 (Peters, 1928); it is not clear whether they make or destroy acetic acid. The most active acetifying organisms were recognised as bacteria by Hansen (1879, 1893) and by Brown (1888), both of whom described additional species. A definitive study of the group was made by Beijerinck in 1899. Dropping the term "Mycoderma," he coined the name "Acetobacter" for them and recognised them as a group with many morphological and physiological similarities.

The most important common property, of course, is the ability to oxidise alcohol to acetic acid. Some make acid from dextrose, but none attack starch, glycogen, or insulin. Many additional species were isolated from malt, beer, wine, and vinegar by Henneberg (1909), and others have been added by Kluyver and his students. Pasteur's two forms are now *A. aceti* and *A. xylinum,* respectively. The Acetobacters are longish gram-negative rods, one or two with a tendency to curve. The surface films on liquid media, which were particularly studied by the early workers (e.g., Hansen, 1895), contain very long chains and filaments up to 400 *p,* in length. The strictly aerobic nature of the group and their gram-negativity suggest relationship to the Pseudomonadaceae, but this is only justified for those members that (when motile) have polar flagella. These are designated *Acetomonas* and may be a single rather variable species (Shimwell and Carr, 1959). They do not have the ability to oxidise acetic acid, while the other members, which when motile have peritrichous flagella, can do so (Leifson, 1954). It is this larger group that is designated *Acetobacter* proper.

Even Acetomonas has important differences from Pseudomonas (Shimwell *et al.*, 1960):

(1) They can form acid on ethanol-agar; some of the Pseudomonads can oxidise ethanol to acetic acid (Stanier, 1947), but only in presence of a growth-supporting medium;

(2) They can oxidise glycerol to dihydroxyacetone;

(3) They can grow at an initial pH of 4.5, as indeed can the Acetobacters too;

(4) They cannot oxidise calcium lactate to calcium carbonate as the Pseudomonads can.

The inability of Pseudomonads to grow at *pH* below 5 or to make surface films distinguishes them from the acetic group as a whole. In addition, of course, they attack proteins and peptides vigorously, and most of them form green, blue, or fluorescent pigments; neither Acetomonas nor the Acetobacters have these properties.

Vinegar Industry

This ancient industry has played an important part in microbiology. Vinegars are made from wines, beers, and ciders by oxidation. In the French process, as carried out at Orleans and elsewhere, wine is stored in broad shallow vats; the acetifying bacteria grow on the surface as a delicate papery skin. Thin pieces of wood may be floated on the wine to help support this membrane, since acetification is most active when the membrane is at the surface; vinegar makers know that it must not be submerged or the process will be slow and the vinegar cloudy. From time to time the vinegar is run off at the bottom of the vat and new wine added, without disturbing the film. The exact species of Acetobacter responsible for the Orleans process is not quite certain. *A. orleanense* was isolated by Henneberg (1909) from such a vat, but the property of forming a coherent film and leaving the liquid clear as in the vinegar vats is not present in this culture.

Other species, such as *A. ascendens,* may cause turbidity, which is hard to get rid of. The Orleans process is said to furnish vinegars of the finest quality. They may contain up to 9.3 per cent (= 1.55 M) acetic acid.

In the "quick" vinegar process, as first developed in Germany, aqueous alcohol or beer is trickled continuously through vats or barrels packed with beech shavings and is converted to vinegar on the way through. A gentle stream of air may be blown through the vat to promote oxidation and to keep the contents cool, since the heat liberated (ca. 113 kcal per mole of acetic acid) is considerable. Here the organisms grow as an almost invisible film on the shavings, and until Pasteur's work it was supposed that the shavings themselves caused the oxidation.

Shavings from actively functioning vessels were used as starter (inoculum) for new vessels. *A. aceti,* isolated by Pasteur and Beijerinck from such vats, is probably the organism of the quick vinegar process. Sometimes its place is taken by the less active mucilaginous or leathery *A. xylinwn,* which continues to grow in the vinegar and may give a bad smell to the product.

Oxidation of Alcohol

The oxidation of alcohol is the most characteristic activity of Acetobacter and Acetomonas. It is formulated as $CH_3CH_2OH + O_2 \longrightarrow CH_3COOH + H_2O$ and mediated (at least in *A. rancens* var. *pasteurianum)* by cytochrome oxidase (Kubowitz and Haas, 1932). As might be expected, it is inhibited by CO_2 in the dark (Tanaka, 1933). There are two steps:

alcohol dehydrogenase

$$C_2H_5OH \longrightarrow CH_3CHO + 2(H)$$

acetaldehyde dehydrogenase

$$CH_3CHO + H_2O \longrightarrow CH_3COOH + 2(H)$$

In very acid media or with concentrated suspensions, acetaldehyde may accumulate (Janke and Kropacsy, 1935). This may be because the aldehyde dehydrogenase functions best in alkaline solution; the enzyme from beef liver has a sharp optimum at *pH*, 8.3 (Racker, 1949). However, the overall optimum for alcohol oxidation is nearer *pH*, 6. The first step (alcohol to aldehyde) can also be brought about by H_2O, through the medium either of catalase (Keilin and Hartree, 1945) or of peroxidase (Wieland and Pistor, 1936). One or other of these oxidases is present in all Acetobacters.

In this process, the hydrogen is carried from one molecule to the other by DPN, which links together the two enzymes, alcohol dehydrogenase and aldehyde dehydrogenase (Racker, 1949). Because the optimum for dismutation is at *pH*, 8.4 or above (Janke and Kropacsy, 1935), it cannot play a significant part in the formation of acetic acid under natural conditions, where the *pH* is usually acid.

With *A. aceti*, the pyruvic acid may be further oxidised to CO_2 + H_2O. Acetate (Visser't Hooft, 1925). A third sequence leads from pyruvic acid to acciylmethyl carbinol; *A. rancens* can give a 25 per cent yield of the carbinol from calcium lactate (Frateur, 1950). (4) *Aldose sugars*, like aldehydes, can be oxidised to acids. Glucose is converted to gluconic acid (Brown, 1886), galactose to galactonic, and arabinose to arabonic acid (Hermann and Neuschul, 1931; Bernhauer and Gorlich, 1935). Some Acetobacters are unable to carry out these reactions, which thus become important for distinguishing species. A number of Pseudomonads, especially *Ps. putida* and *Ps.*

fluorescens, oxidise glucose, arabinose, xylose, and ribose in the same way (Lockwood and Nelson, 1946). In the case of glucose, if the further oxidation of gluconic acid is prevented, the oxidation can be almost quantitative (Entner and Stanier, 1951).

The first product of oxidation is the lactone. *Ps. saccharophila* converts arabinose rapidly to arabonoy-lactone, DPN being hydrogen acceptor (Palleroni and Doudoroff, 1957). Similarly *Acetomonas oxydans* converts glucose to glucono-8-lactone (King and Cheldelin, 1957-1958), and in this case two enzymes are present: one is in the particles that contain the cytochromes and has optimum *pR*, 5.5, the other is soluble and has optimum *pH* 8.6. The cells also contain an enzyme to hydrolyse the lactone to gluconic acid. Evidently they are equipped to carry out the oxidation under a variety of conditions.

Often the resulting acid, which contains secondary carbinol groups, undergoes a second stage of oxidation as in (3) above, yielding a ketonic acid.

The 5-keto acid is formed by many Aretnfcprters and often in good yield (Bernhauer and Gorlich, 1935); innprMa wit) gait often crystallises out characteristically.

The 2-keto acid is also formed by many Acetobacters, though some times only in traces (Frateur, Simonart, and Coulon, 1954). It is formed more actively by many Pseudomonas species, yields as high as 80 per cent having been recorded (Lockwood *et al.*, 1941). The acid is of especial interest because on decarboxylation it yields arabinose, which may thus be derived under natural conditions from glucose.

In fact, the almost constant association of glucose and arabinose in cell-wall materials of higher plants suggests that this path of breakdown may be of wider occurrence than in Acetobacter alone. The conversion of 6-phosphogluconic acid to ribulose and ribose is a comparable reaction. The 2-keto acid is further oxidised, by pigment-forming strains of Acetomonas, to the 2,5-diketo acid; this is the precursor of a characteristic brown pigment (Beijerinck, 191 la; Katznelson *et al.*, 1953), which gave the name *A. melanogenum* to these strains; it occurs also in two strains of *A. aceti* (Carr and Shimwell, 1960). Glucose and gluconic acid yield the same product. Slow further oxidation, at least with *Ps. fluorcscens* leads eventually to α -ketoglutaric acid (Lockwood and Stodola, 1946), though some cultures can oxidise glucose without being able to oxidise the 2-keto acid (Entner and Stanier, 1951). Since ketoglutaric acid is also the end product of the oxidation of aromatic and hydroaromatic compounds, it seems likely that (either directly or after amination to glutamic acid) it is assimilated readily by the cells.

The clarity of the sugar oxidations was greatly enhanced by the discovery of "Acetobacter suboxydans," now *Acetomonas oxydans*, an organism that does not "overoxidise" the products (Kluyver and de Leeuw, 1924; Visser t'Hooft, 1925). Not only does this bacterium not oxidise acetic acid at all, but it does not oxidise 5-ketogluconic acid, so that in four-day-old cultures its conversion of glucose to 5-ketogluconic acid is virtually 100 per cent (Kluyver and Boezaardt, 1938). That the lactone is intermediate was noted above. Gluconic acid is similarly converted.

A 2-keto acid is also formed from arabonic acid by *Ps. saccharophila* (Palleroni and Doudoroff, 1957). This is not an oxidation but a dehydration like that of phosphogluconic acid caused by this and other pseudomonads. The resulting 2-keto-3-deoxy-arabonic acid splits to yield pyruvic and glycolic acids, and the whole reaction may well be more general, both as to the substrates attacked and the organisms carrying it out. Some Acetobacters contain many of the enzymes active in glucose fermentation. Thus, *A. suboxydans* and *A. rancens* var. *pasteurianum* could convert hexosediphosphate to triose, acetaldehyde, and acetate (Simon, 1930). These organisms, therefore, have all the enzymic outfit needed for conversion of glucose to vinegar. Anaerobically, in heavy suspensions, they can ferment glucose to alcohol and CO_2, an unusual fermentation for bacteria.

Oxidations of Sugar Alcohols: Sorbose Bacterium

The berries of the rowan tree or European mountain ash, *Sorbus aucuparia,* yield the sugar alcohol sorbitol. Because certain samples of the berry juice yielded also the rare keto sugar *sorbose,* which is not present in the fresh juice, Bertrand (1904) subjected the juice to an extensive study, which turned out to be most illuminating. The fresh juice contained no sorbose, but did contain glucose (4.4 per cent). On setting aside, it underwent spontaneous fermentation by a yeast, which consumed most of the glucose, but formed no sorbose. This was followed by growth of a fine pellicle of "Mycoderma" which oxidised the alcohol formed by the yeast, and also by a Penicillium; neither of these, however, produced sorbose. Finally he observed that a Drosophila visited the solution and laid eggs; almost immediately afterwards a pellicle of a different type appeared gelatinous and consistent, becoming greenish as it dried. Now the solution was highly reducing and contained quantities of sorbose. Examination of the new pellicle showed it to be an almost pure culture of a rod-shaped bacterium, subsequently identified with *A. xylinum.* Bertrand found that the bacterium gave an 80 per cent yield of sorbose from sorbitol and that it also formed reducing sugars from other alcohols in yeast extract medium, including mannitol, erythritol.

It will be noted that the formation of 5-ketogluconic acid is in accordance with the rule; 2-ketogluconic acid is, however, an exception. There are other exceptions too, especially outside the acetic bacteria; a Pseudomonas can form dehydrogenases for dulcitol and iditol (both with the terminal OH groups *trans* to one another), and *E. coli* forms a similar enzyme for dulcitol-1-phosphate (Shaw, 1956; Wolff and Kaplan, 1956).

Cell-free enzymes have been obtained for many of these oxidations. *Acetomonas oxydans* yields two different mannitol dehydrogenases, one active at *pH. 5* and linked to cytochrome in particles, the other with optimum at *pH,* 8, soluble and reducing DPN (Arcus and Edson, 1956; cf. the two glucose dehydrogenases above). *E. coli* apparently carries out the same oxidations *via* the phosphates, because when it is grown on the alcohols, enzymes reducing DPN with mannitol-1-phosphate, sorbitol-6-phosphate, and dulcitol-1-phosphate are developed (Wolff and Kaplan, 1956).

The first-named, which forms fructose-6- phosphate, was purified; it is probably the basis for the utilisation of mannitol by Streptococci and Pneumococci also. The conversion of sorbitol to sorbose in high yield has become of importance in the industrial manufacture of ascorbic acid (vitamin C), which is prepared by oxidising sorbose.

Since naturally occurring sorbose is rare, its production from sorbitol is essential. Besides *A. xylinum,* Acetomonas strains also give good yields of sorbose (Bernhauer and Gorlich, 1935; Boeseken and Leefers, 1935). The D-forms, but not the L-forms, of many sugar alcohols are oxidised by Acetomonas according to Bertrand's rule (Hann *et al.*, 1938), and it is evident that the rule is an expression of a reaction mechanism characteristic for the acetic bacteria but not for other groups.

Physiological Grouping of Acetobacter Types

As with other bacteria, the same Acetobacter has often been isolated under different names, while different cultures have sometimes been given the same name. For example, *A. aceti* Hansen is not the same as *A. aceti* Beijerinck and two organisms named *A. ascendens* have quite different properties. Many of the characters described are not permanent; for example, "*A. pasteurianum*" contained a starchlike compound staining blue with iodine, but this seems to be lost on continued culture, and the resulting organism is then indistinguishable from *A. ranccns* (Frateur, 1950).

Apart from minor morphological differences and flagellation, it is the biochemical characters that are most characteristic. These may vary slightly with age, for *Acetomonas oxydans* in very young cultures oxidises glucose to CO_2, but in older cultures converts it only to 5- ketogluconic acid (Butlin, 1938; Kluyver and Boezaardt, 1938). In general, however, the biochemical properties are fairly stable. The yields of various acids, ketones, sorbose, and so on, obtained after three to four months' incubation, show a continuous series of increasing oxidising power from *A. gluconicum* (probably a variety of *Acetomonas)* to *A. ascendens* (Hermann and Neuschul, 1931).

A much more extensive series has been worked out by Frateur (1950), using Kluyver's extensive collection of Acetobacters. This makes possible a classification based primarily on the oxidising power. The most oxidising species do not form ketones or gluconic acid, because the oxidations go to CO_2 and water; the least oxidising species, which give good yields of ketones and gluconic acid, cannot attack acetate or lactate. Some strains of the Oxydans group form a little 2-ketogluconic acid (Frateur *et al.*, 1954).

The absence of catalase in the Peroxydans group does not mean a flavoprotein type of respiration as in the Lactobacilli, because the respiration is inhibited by cyanide and azide, and there is no detectable accumulation of H_2O_2. Probably the oxidising enzyme is cytochrome peroxidase (Lenhoff and Kaplan, 1956). The ability to grow on a mineral medium, with nitrogen derived only from NH_4^+ and carbon derived only from alcohol, is a valuable criterion, especially for *A. aceti,* but has of course no relation to the oxidations; it relates only to the synthetic abilities of the organisms. Many Acetobacters require preformed amino acids and vitamins in the medium, while some, like *A. melanogenum,* can form all needed amino acids from ammonium, but require several vitamins (Foda and Vaughn, 1953).

Formation of Cellulose

One of the striking properties of these bacteria is the ability to produce cellulose. Cellulose formation is limited to *A. xylinum* and the related *A. acetigenum* (which Frateur considers a variety of *xylinum).* The cellulose fibrils, which do not differ in any major respect from the fibrils formed in plant cell walls, are not in the bacterial wall but outside it. They form a tangled mass at the surface of the medium. In some media, they become impregnated with nitrogenous material and, consequently, when tannic acid is added, they form a heavy insoluble mass; in World War I it was even proposed to make "bacterial leather" in this way. In order to deposit the insoluble cellulose fibrils outside the cell, a soluble precursor that can polymerise has to be excreted.

When the cells take up C^{14}-glucose, the isotope is transferred to the cellulose. This shows that polymerisation of glucose units is the source of cellulose. But the precursor is not an oligosaccharide nor a sugar phosphate, although these are excreted to some extent (Walker and Wright, 1957; Ziegler and Weigl, 1959). Instead, extracts of *Acetobacter xylinum*, on mixing with an ultrafiltrate from the cells, deposit microfibrils insoluble in alkali and apparently true cellulose (Colvin, 1959; Khan and Colvin, 1961). The active extracts are *soluble in alcohol* and contain a complex of glucose and an aliphatic alcohol; the ultrafiltrate contains an enzyme. *A. acetigenum* behaves similarly (Brown and Gascoigne, 1960). Thus the cells take up glucose, combine it with lipid to form precursor, probably at the cell membrane, and excrete it together with an enzyme that removes the hexose residues and orients them into fibrils. This ready excretion may be connected with the presence of small modified areas on the cell surface (Webb and Colvin, 1962, 1963). The lipid is then reabsorbed.

The whole remarkable phenomenon may be a model for the excretion of polysaccharide and polyuronide capsules, as well as the sheaths of Chlamydobacteria and Myxophyceae.

Oxidations of Aromatic and Hydroaromatic Compounds

Under this head, oxidations brought about by Acetobacter, Pseudomonas, Vibrio, and certain other organisms will be treated together. It was first shown by Beijerinck (1900) that an Actinomycete (obtained from tree roots) produces a brown pigment on gelatin media and at the same time hardens the gelatin and makes it insoluble in hot water. This phenomenon was traced to the formation of benzo-quinone by the organism as an oxidation product of an amino acid. Evidently the Actinomycete is oxidising the side-chain of aromatic amino acids and at the same introducing extra OH groups into the benzene ring to form either catechol or quinol, which is then oxidised to the ortho or para-quinone:

> Similar steps occur in the formation of melanin from tyrosine, although this Actinomycete does not attack tyrosine. It also cannot attack the final quinone, which accumulates.

A more powerful oxidation is exerted by *Pseudomonas fluorescens,* which oxidises several aromatic acids to straight-chain aliphatic compounds. This action is inducible; cells grown on yeast extract and suspended in the aromatic acid take up oxygen at only the basal rate for a time, but then suddenly increase their respiration.

If grown on the acid as sole carbon source, they are "preinduced" and oxidise it without any delay. The induction can be prevented by antibiotics, or hastened by supplying ammonium and succinate (Bernheim and do Turk, 1953). Evidently it does not differ from induced enzyme formation in general. The phenomenon was used by Stanier to study the pathway of oxidation. For example, if mandelic acid is oxidised by way of benzoic acid, then cells preinduced to oxidise mandelate must also be preinduced to benzoate, but not vice versa.

Further study of the same sort has shown that benzoic acid, phenol, and also anthranilic acid are all oxidised to catechol. If benzoic acid isotope-labelled in the carboxyl is used, unlabelled catechol results, but benzoic acid labelled at C-1 gives rise to catechol labelled at C-1 and C-2 (Sleeper, 1951). The carboxyl group is thus evidently replaced by OH, and then a second OH inserted (Parr *et al.,* 1949): However, this is not the only route, since in another organism salicylic acid is an intermediate (Bhat *et al.,* 1959). Catechol in turn is oxidised by two different routes. The first, by a pair of enzymes extracted from *Ps. fluorescens* (Hayaishi and Hashimoto, 1950; Evans and Smith, 1951) or *Vibrio OI* (Parr *et al.*, 1949), splits the ring *between* the two OH groups to form *cis-cis-* muconic acid and thence, 2-keto-adipic acid (which finally yields succinate and acetyl-CoA). The second, by an enzyme obtained from two Pseudomonads, splits the ring bond *adjacent* to one of the OH groups, to form α -hydroxy-muconic semi-aldehyde and thence pyruvate.

The latter organisms do not attack *cis-cis-muconic* acid. Protocatechuic acid, which is 4-carboxy-catechol, yields the carboxylated analogues in each case; some Pseudomonads produce 0- carboxy-m-czV-muconic acid (MacDonald *et al.*, 1954; Gross *et al.*, 1956), while others form the y-carboxy derivative of a-hydroxy- muconic semi-aldehyde (Dagley *et al.*, 1960). The conversion of protocatechuic to oxalic acid by *Asp. niger* (Bernhauer and Waelsch, 1932) suggests the latter route. Meta- and para-hydroxy- benzoic acids follow the first route, yielding 3-keto-adipic acid; each first adds a hydroxyl to form protocatechuic acid, which is further oxidised as above. Phenylacetic acid lies on a separate pathway, because cells that can oxidise it require a further induction period to oxidise benzoate (Dagley *et al.*, 1952). Again there are two routes, both taken by strains of *Ps. fluorescens;* one goes via the 3,4- and the other by the 2,5- dihydroxy acid (Kunita, 1955). The latter is homogentisic acid.

Since mandelic acid is oxidised to benzoic acid, probably via benzoylformic acid and benzaldehyde, and since anthranilic acid can be formed by the oxidation of tryptophan, a large number of oxidation pathways for aromatic compounds are known. The muconic semi-aldehyde routes are omitted, for clarity. Tyrosine and phenylalanine lie on a different pathway (Stanier, 1950) since tyrosine is oxidised first to DOPA and thence to 2,5-dihydroxy-phenylacetic, or homogentisic acid. This is split to maleyl-acetoacetic and finally yields fumarate and acetoacetate. *Ps. fluorescens* and Vibrio OI when adapted to oxidise tyrosine also oxidise homogentisate, and young cultures accumulate the latter (Dagley *et al.*, 1952; Chapman and Dagley, 1960). Gentisic acid, with one CH_2 group less, behaves similarly, yielding maleyl-pyruvate and hence fumarate and pyruvate (Lack, 1959; Dagley *et al.*, 1960). These reactions of geritisic acid are important because it is itself an oxidation product of salicylic and metahydroxy benzoic acids by *Ps. fluorescens.*

In general, these oxidations follow recognisable patterns; introduction of OH groups, splitting the benzene ring in two ways as noted above, and stepwise oxidation and decarboxylation of the aliphatic products. Many of the oxidases have been prepared cell-free, and in some cases work with bacteria has been paralleled by feeding experiments with animals.

A few Pseudomonads and many Mycobacteria can oxidise the aromatic hydrocarbons themselves, in spite of their insolubility in water, as well as the phenols, which are normally very toxic (Tausson, 1927; Gray and Thornton, 1928). Application of the sequential induction technique to the attack of naphthalene by five of these has shown that when grown on succinate, they oxidised naphthalene only after a lag, but when grown on naphthalene, they all oxidised it immediately (Treccani *et al.*, 1954).

Growth on aliphatic hydrocarbons, by contrast, is mainly limited to the Mycobacteria and Nocardias. In these cases the study of phenyl-substituted hydrocarbons has made clear that the mode of attack is by repeated oxidation (Webley *et al.*, 1956). However, *Ps. aeruginosa* can dehydrogenate n-heptane to hepta- 1-ene, using DPN as H acceptor (Chouteau *et al.*, 1962). The complex hydrocarbon structure of rubber can also be oxidatively attacked, mainly by fungi and Streptomycetes (Rook, 1955). Many hydroaromatic compounds are also oxidised. These compounds are derivatives of cyclohexane and include myo-inositol. Some of them act as intermediates in the formation of aromatic compounds, to which they are readily oxidised. Thus, quinic acid, $C_3H_7(OH)_4COOH$, is oxidised to protocatechuic acid or catechol by *Ps. fluorescens, Ps. calcoacetica,* and *Ps. aromatica* (Beijerinck, 191 Ib), *A. gluconicum,* and several fungi (Bernhauer and Waelsch, 1932; Bernhauer and Gorlich, 1935) 5 quercitol is oxidised to pyrogallol.

Inositols are also oxidised to monoarid di-ketones, apparently following rules suggestive of Bertrand's rule. The halophihc organism, *Ps. beijerinckii,* isolated from salted beans (12 per cent NaCl), turns the beans bright purple (Hof, 1935). The pigment was produced only on a medium containing aqueous bean extract, the essential component of which turned out to be myoinositol (Kluyver *et al.* 1939). Although not fully proved, the purple pigment is probably the Ca salt of tetrahydroxy-quinone.

Bacterial Synthesis of Aromatic Compounds

A parallel development of recent years has been the understanding of how bacteria synthesise aromatic compounds. This has benefited from the work on hydroaromatics, but mainly developed from the finding that a number of mutants of *E. coli* require tyrosine, phenylalanine, tryptophan, p-aminobenzoate or p-hydroxybenzoate (Davis, 1950).

Some require more than one such compound for normal growth. These strains evidently fail to form the benzene ring, and in several cases they accumulate significant intermediate compounds in the medium; the identification of these and study of their metabolism by other mutants (largely due to B. D. Davis and his co-workers) have made it possible to construct the route of biosynthesis of the benzene ring in bacteria. Since these compounds contain 7 carbon atoms they obviously are not formed directly from hexose, but they arise from the 7-carbon sugar, sedoheptulose. Sedoheptulose-7-phosphate and the 1,7-diphosphate were slowly converted to shikimic acid by a cell-free system prepared from the *E. coli* mutant, 83-2, which accumulates this acid. However, a mixture of the 4-carbon and 3-carbon compounds, erythrose-4-

phosphate and phospho-enol-pyruvate, was much more effective, giving 88 per cent conversion to dehydroshikimic acid in one hour. Closer study of this system with purified enzymes showed that the initial condensation product was a new 7-carbon acid, 2-keto,3- deoxy-D-araboheptonic acid 7-phosphate, and that this then cyclises to 5-dehydroquinic acid, (Srinivasan *et al.*, 1959):

> This, then, is the elusive reaction in which the 6-carbon ring is first formed. The condensation is shown as due to a hydrolysis, the phosphate radical of pyruvic acid attracting a hydroxyl, which drives the electrons towards the carbon, liberating the phosphate and setting up a bond to the carbon-1 of the erythrose, which has combined with the proton. The formation of phosphoenolpyruvic acid from the first 3 C-atoms of the heptulose proceeds.
>
> Quinic acid is interconverted with its 5-dehydro-form by a dehydrogenase present in Aerobacter. This acid does not lie on the main pathway, although it occurs fairly widely in plants. We have come a long way from the simple oxidation of alcohol to acetic acid and have moved away from the acetic bacteria to the coliforms, but the path of research in this area, though perhaps a winding one, has been continuous and logical.

Breakdown of Disaccharides and Polysaccharides Breakdown of Disaccharides

The Disaecharidases: The disaccharides are in general attacked by bacteria only after hydrolysis monosaccharides. The main exception to this rule occurs in the oxidation of sugars toonic acids. Since this bacterial oxidation involves only one end of the sugar molecule, it is perhaps not surprising that it can lake place when a carbon atom farther down the molecule is attached to a second saccharide. For example, maltose, which is 4-α -glucosidoglucose, is oxidised to maltobionic acid, and lactose to lactobionic acid, by certain Pseudomonads *(Ps. quercito- pyrogallica* and related species; Kluyver *et al.*, 1950).

The fermentations, however, involve breakdown of the monosaccharide structure and require that the hexose (or pen lose) be free. Hence, if a disaccharide is riot hydrolysed, it cannot usually be fermented.

For this reason the ability to hydrolyse disaccharides is generally measured by their fermentation. Because this ability is highly characteristic for different organisms, it has been widely applied in the taxonomy of bacteria and fungi as well as in the bioassay of disaccharides. The coliform bacteria are divided on the basis of lactose fermentation, Escherichia and Aerobacter attacking lactose, Proteus and Salmonella not. All four attack maltose. The increasing knowledge of bacterial genetics and of the control of enzyme formation makes these separations less satisfactory.

At one time the hydrolysing enzymes or disaccharidases of different organisms were used for the quantitative analysis of sugar mixtures. By determining the reducing power of the solution before and after fermentation by a series of species of Torulopsis, Monilia, and Saccharomyces, the amounts of glucose, fructose (or mannose), galactose, sucrose, and maltose could be determined (Kluyver and Hoogerheide, 1933; Harding and Nicholson, 1933; van Voorst, 1938-1939). These methods have now been superseded by chromatography; by suitable choice of solvents, the sugars can be cleanly separated as spots on paper, revealed by spraying with ammoniacal $AgNO_3$ or aniline, or eluted and the reducing power determined. The galactosidase of *E. coli* has been more extensively purified than any other such enzyme (probably to 98 per cent purity) and found very stable and specific (Cohn, 1957). It has a molecular weight of about 700,000, and 1 mole hydrolyses some 4000 moles of the test substrate (o-nitrophenyl-yft-galactoside) per second, a relatively high activity. Moreover, it not only hydrolyses galactosides but also transfers galactose residues to other acceptors, forming trisaccharides, etc. The specificity varies somewhat with the monovalent ion used to activate the enzyme, activity on lactose being highest with K^+ and on nitrophenylgalactoside highest with Na^+. Apparently the enzymes of several *E. coli* strains, of Aerobacter, and Shigella are all the same, while those of yeast and Lactobacillus differ both immunologically and in their affinity to inhibitors. Other disaccharidases are less well studied.

The α -galactosidase of *E. coli* resembles its counterpart in that its formation is induced by substrates, like melibiose and raffinose, but it cannot yet be obtained cell-free (Sheinin and Crocker, 1961), although it can be prepared from yeast. It is not highly specific in its substrates.

Enzymes hydrolysing maltose and other α -glucosides (hence α -glucosidases) are readily obtained from brewer's yeast or *Aspergillus oryzae* by autolysing. They have their optimum *pH* at 4.5, while the bacterial enzymes act optimally at 6.5 to 7; differences in specificity also make it probable that these enzymes are not the same. Invertase again has been mainly studied from yeast (Hestrin *et al.*, 1955); it hydrolyses sucrose, the fructoside linkage in raffinose, and some fructose polymers (inulins and levans) but does not attack galactosides or glucosides. Like 0-galactosidase, it can transfer fructose residues to other sugars and alcohols.

Induced Formation of the Enzymes

The disaccharidase content of an organism commonly varies in amount according to the medium used for culture. As long ago as 1899, Duclaux found that a strain of *Aspergillus niger* produced invertase when it was grown on a sucrose medium. Similarly, Aspergillus grown on a milk medium could hydrolyse casein, and yeast grown on a galactose medium could ferment galactose (Dienert, 1901). These phenomena were investigated systematically by Karstrom (1931, 1938), who found that certain enzymes were always present, irrespective of the medium, while others were formed only when the medium contained their substrate. Glucose is fermented in all cases, but fermentation of the other sugars is in large part dependent on their presence in the growth medium. Note that growth in lactose permits the fermentation of galactose, but not vice versa; i.e., lactose leads to formation of the galactose-fermenting system *galactozymase* which has three constituents, an enzyme, uridine-diphosphoglucose as co-enzyme, and DPN as hydrogen transferring agent (Kalekar and Maxwell, 1958).

However, galactose does not lead to formation of the lactose-hydrolysing enzyme galactosidase. *E. coli* gave similar results and so have several yeasts (Kluyver and Custers, 1940). Karstrom termed the enzymes that are always present "constitutive," and those that are formed only in the presence of the substrate "adaptive." The distinction is sometimes quantitative rather than qualitative, since cells grown in the absence of sucrose do hydrolyse sucrose slowly. Also, hydrolysis even of lactose may begin after a long induction time; this time may be that needed to synthesise the enzyme. The term "adaptive" has since been replaced by "induced." Induction, is widespread and appears in the metabolism of almost every group of substances; indeed, there is reason to believe that in the bacteria all enzymes are interrelated with the induction process even when their formation is constitutive. Sometimes their formation can be induced not only by the substrate but also by a related compound. Genetic analysis has shown in several cases that the formation of an enzyme is controlled by a pair of genes at the same locus, i.e., "alleles," and that *inducible* is dominant over *constitutive.*

In the present connection the inducible enzymes include not only the disaccharidases but also those attacking monosaccharides, such as the phosphorylase or kinase involved in the initial phosphorylation, or systems like that required for the galactose-glucose conversion in galactose-grown yeast. For example, *E. coli,* when grown on xylose, ferments xylose but not arabinose, and when grown on arabinose, it ferments arabinose but not xylose. Grown on glucose it ferments neither (Karstrom, 1931).

Sucrose occupies a special position since it need not be hydrolysed, but can be phosphorylised. The enzyme *sucrose phosphorylase,* obtained from *Ps. saccharophila,* establishes equilibrium between glucose- 1-phosphate and fructose, and sucrose and phosphate. Hence *Ps. saccharophila,* in which this enzyme is inducible, can synthesise sucrose from the mixture. It can also grow on sucrose or on trehalose and exhibits the curious behaviour of oxidising these faster than their constituent hexoses (Doudoroff, 1940). Similarly in *Ps. putrifleans* the induced cells not merely oxidise but also *grow* faster on sucrose than on hexose (Doudoroff *et al.,* 1949). The reason in both cases is that the phosphorylative split occurs more readily than does direct phosphorylation of hexose. *Neisseria meningitidis* contains a maltose phosphorylase which acts similarly, forming glucose and glucose-1-phosphate (Fitting and Doudoroff, 1952).

Two phenomena are involved in adaptation; they may be considered as "Darwinian" and "Lamarckian." The former is that of *selection.* If, out of the whole population in a culture, one or two cells mutate and begin to produce the enzyme for attacking the disaccharide, then when the disaccharide is added to the medium, these cells are greatly favoured and multiply faster than the others. Suppose, for instance, that the favoured cells are in a minority of one in a million, but that they divide every 20 minutes, while the others divide only once in an hour. Then we have:

Favoured Others

The selection process undoubtedly underlies many of the earlier cases of adaptation. It seldom require several transfers before the organisms are "trained" to utilise the new substrate. It requires, of course, that the organisms produce, in small numbers, mutant cells containing the new enzyme. This interpretation has been worked out in the classic case of *"B. coli mutabile"* a lactose-negative strain of *E. coli,* in which Neisser

(1906) and Massini observed that old colonies on lactose-peptone agar produced daughter colonies, or "papillae," which could ferment lactose.

The change was detected because the medium (Endo's agar) contained fuchsin sulphite; this turns deep pink in the presence of aldehydes and thus indicates whether the sugar is being fermented. In this case, the white colonies, growing only on the peptone, developed pink papillae. The explanation is that the strain produces in every generation about two cells in 10^7 that can ferment lactose (Lewis, 1934; Ryan, 1952). Presence of lactose in the medium naturally allows these to overgrow the others, and in a short time the whole culture is lactose-positive.

Even in the absence of lactose, the bacteria can grow sufficiently on traces of impurities so that lactose-fermenting mutants can be produced (Ryan, 1952). The rate of mutation is not affected by the presence of lactose, nor indeed of any of a variety of related glucosides, but depends strictly on chance. Mutation can of course occur in the opposite direction, producing lactose-negative strains; in *E. coli* K_{12} many different lactose-negative strains have been identified.

Adaptation to ferment lactose can also occur without cell multiplication. In these cases a new enzyme is being formed in the cells as a response to the substrate. This "Lamarckian" phenomenon is induction proper. It has been studied on enzymes of many types in addition to the disaccharidases.

Action of Bacteria on Starch

Starch is attacked by so many organisms, both bacteria and fungi, that there can hardly be said to be a special microflora of starchy products. There are characteristic differences between products, however, as exemplified by Beijerinck's observation that certain samples of barley flour regularly underwent butanol fermentation, while others gave only butyric organisms, and still others only *Bac. polymyxa*. All these bacteria hydrolyse starch actively; so also do many other sporeformers, both aerobic and anaerobic, and many fungi, but very few yeasts. The ready attack on starch is probably due to its similarity to glycogen, which is a true storage and reserve product in many bacteria.

Understanding of the nature of starch is essential to an understanding of its attack by microorganisms. Its structure depends on the following facts (Myrback, 1948):

1. Starch contains two constituents; *amylose,* a straight chain polymer containing about 200 to 300 glucose units, and *amylopectin,* a branched structure of molecular weight greater than that of amylose and containing phosphate groups. Similar to amylopectin is the glycogen of animal tissue and many plant starches which turn red with iodine, as, for instance, that in Oscillatoria (Hough *et al.*, 1952).
2. When starch is hydrolysed by enzymes, maltose and glucose are the only reducing sugars produced. Since glucose is produced from maltose by hydrolysis, it is deduced that both amylose and amylopectin consist of large numbers of maltose units linked together.
3. When starch is fully methylated, so that all free hydroxy groups have reacted, and is then hydrolysed, 2, 3, 6-trimethyl glucose is nearly the entire product. It follows that almost all the molecules are linked through the 1-4 positions. Amylopectin, however, yields a little 2,3-dimethylglucose, which shows that a few of its hexose units are linked at both the 4 and the 6 positions. These introduce "branch points" into the chain. Also, in both constituents, the few terminal units yield a little tetramethyl glucose. Hence amylose must be a polymer of 1:4-linked a-glucose units, and amylopectin the same with occasional 1-6 links in addition. Starch-hydrolysing enzymes or amylases are of two kinds. The commonest, called a-amylase, causes rapid disappearance of the iodine colour with only very small liberation of reducing groups. Starch gels are rapidly liquefied, and at the

completion of the hydrolysis low molecular-weight "dextrins" remain. With Taka-amylase (from *Aspergillus oryzae)*, these dextrins contain from 3 to 7 hexose units. If pure amylose is the substrate, however, it is completely hydrolysed. The other enzyme, amylase, rapidly sets free reducing sugar but does not destroy the iodine colour. Hydrolysis stops when 60 per cent of the possible maltose is formed, and the remaining material, called "limit dextrin," has a very high molecular weight, about half that of the original starch. Again, pure amylose is hydrolysed completely.

It is clear that amylase breaks off maltose molecules from the "outside" of the amylopectin molecule, leaving a good part of it intact, while α -amylase attacks interior linkages, breaking up the molecule into fragments of fairly large size. Its action is stopped by the branch-points, where a glucose residue is linked at both the 4 and 6 positions. The average length of the branches is about 25 glucose residues and the type of branching is probably bushlike and multiple, rather than short branches on a long straight chain.

The amylases of bacteria and fungi are of the *a* type. The enzyme from *Bac. subtilis* resembles that of malt, both in reaction kinetics (rapid fall of viscosity, slow liberation of reducing sugar) and in the nature of residual dextrins (Hopkins *et al.*, 1942; Bernfeld and Fuld, 1948).

The amylase of *Cl. acetobutylicum*, which breaks down starch completely to a 100 per cent yield of maltose (Hockenhull and Herbert, 1945), is probably a mixture. The amylase from *Bac. subtilis*, which has been crystallised (Meyer *et al.*, 1947), requires free SH groups for its activity and is therefore, inhibited by heavy metals; it also requires Ca ions (Di Carlo and Redfern, 1947-1948). Many streptococci contain a starch, which hydrolyses completely to maltose with pure *a*- or J-amylase, and is therefore, a straight-chain amylose (Crowley and Jevons, 1955; Baum and Crowley, 1960); these organisms form amylases too. The formation of bacterial a-amylase is readily induced by suspending cells of *Ps. saccharophila* in a starch medium (Markovitz and Klein, 1955) and, being an extracellular enzyme, this makes a favourable system for studying enzyme induction. There is in the same bacteria an *intracellular* -amylase.

The action of *Bac. macerans* on starch yields very peculiar dextrins, known as Schardinger dextrins. These consist of rings of 6, 7, or 8 glucose units (i.e., molecular weight about 1000), linked 1-4; they are non-reducing, are resistant to 2-amylase, and give characteristic crystals with iodine (French, 1957). They are not synthesised from maltose, but are formed direct from amylose, and appear when the average chain length has fallen to about 100 glucose units. Their formation has been considered as evidence that the amylose chain is a spiral in which one complete turn occupies 6-8 hexose units, but it is more likely that the *macerans-enzyme* twists the starch molecule into spirals (Myrback and Neumuller, 1950); in either case it appears that as the dextrins are hydrolysed off from the parent molecule, their ends at once recombine. The purified enzyme, *dextrinogenase,* is free from true amylase (Schwimmer, 1953); it forms 43,000 moles of dextrin per mole of enzyme per minute.

With the increasing interest in amylase for industrial use (destarching of textiles, preparation of potato and grain for fermentations), some attention has been given to bacilli as a source. The most active amylase-formers are strains of *Bac. subtilis, mesentericus, macerans* and *polymyxa*. The enzyme, which is freely excreted into the liquid medium, generally along with proteases, is produced best in shallow cultures or with vigorous aeration; in other words, enzyme formation is linked to oxidative metabolism. Since amylase formation is inducible, it is produced in largest amounts on a starchy substrate such as potato or cereal mash.

Hydrolysis of Pectins

The bulk of the organic material of plants is made up of the polysaccharides, polyuronides, and aromatic material of the plant cell wall. Because of the importance of these substances in the world's cycle of carbon, their breakdown will be considered here systematically. The pectins form the "intercellular substance" of higher plants, linking the cellulose walls of one cell to those of the adjacent one.

They also occur in water-soluble form. They are essentially methylated polymers of D-galacturonic acid; the demethylated product is called pectic acid. Thus they are not polysaccharides, but polyuronides. Most of the preparations contain small amounts of galactan and araban (i.e., polymers of galactose and arabinose, respectively) in some kind of ester linkage.

Pectins are attacked by numerous organisms, especially *Enterobacteriaceae* and the sporeformers. One of the main groups of pectindecomposers is the soft-rot organisms. These bacteria grow in the middle lamellae of many tubers, leaves, and stems, causing the tissue to break up into a soft mass.

In the early stages, intact loose plant cells can be seen, but later these are killed. One such organism was first isolated by Jones (1901, 1905) from carrots, under the name of *Bacillus carotovorus,* later changed to *Erwinia carotovorum.* It is highly non-specific and is a common pest in commercial vegetables such as cabbage, beets, and celery. The rapid spread of cellular breakdown convinced Jones that an exoenzyme was produced, and though cell-free filtrates were at first inactive, the presence of exocellular enzyme was shown by separating the bacteria from the plant tissue by a layer of agar.

These bacteria are gram-negative rods, carrying out a typically formic fermentation of sugars or pectins. Except for the ability to attack pectin, and therefore, to macerate plant tissue, the Erwinias could be assigned to the genera Proteus and Aerobacter (Elrod, 1942). Besides them, some 5 per cent of a large group of soil Pseudomonads could liquefy pectate gel and macerate cucumber tissue (Paton, 1958). Pectin is also decomposed by the Aerobacilli and numerous clostridia, the latter being the main organisms causing the "retting" of flax, hemp, and jute.

Retting is the removal of carbohydrates other than cellulose from the stems of the fibre plants. The method, which until the last 50 years was essentially the same as that used in ancient Egypt and by the European lake dwellers, consists simply of immersing the flax (or hemp) stalks in small ponds for several days. Coconut fibre is obtained in a similar way (Heyn, 1951). The changes occur in three stages:

1. In the first 6 to 12 hours (at 20° C) the tissue swells and becomes waterlogged; some of the soluble matter diffuses out.
2. For about four days aerobes, including lactic bacteria and yeasts, develop, fermenting the solutes and using up the dissolved oxygen.
3. Thereafter clostridia develop, hydrolysing the starch and pectin and carrying out a butyric fermentation on the products. As a result the parenchyma breaks up, and the vascular bundles with their strong bast fibres (linen) can be separated. The end point comes when the total acids reach a constant value. From this point on, celluloseattacking clostridia may cause damage to the fibres. Attempts have been made, especially in Italy, to use known enrichment cultures as "starters" in the retting pit. *Cl. felsineum* has given the best results. Aerobes, including *Bac. comesii*, also ret very efficiently, but under aerobic conditions (obtained by blowing air through the pit or tank) cellulose-destroying organisms may develop rapidly and begin degrading the fibres.

The breakdown of pectins requires at least three enzymes. The first stage, conversion of insoluble protopectin to pectin, is what causes separation of the plant cells or *maceration*. The enzyme causing this, protopectinase, has been found in many Aspergilli, and in *Cl. felsineum* (Heyn, 1951). The second stage is the demethylation of pectin to pectic acid; this enzyme, called pectin methyl esterase, is widely present in bacteria and higher plants. The final stage, hydrolysis of the 1-4 linkages to break up the macromolecule, is due to a third enzyme, pectinase, obtainable from cultures of *Asp. oryzae, Rhizopus tritici,* and *Erwinia carotovorum.* Its *pH* optimum is near 3. From its products, yields as high as 85 per cent of crystalline D-galacturonic acid have been obtained (Pigman, 1957).

Hydrolysis of other Natural Polysaccharides

Plant Cell Walls Contain Various Insoluble and Fibrous Constituents: Cellulose is the best known of these, but in addition many other polysaccharides, called *hemicelluloses,* are soluble in dilute alkali and hydrolysed by dilute acids. The commonest are the xylans, which constitute about half the weight of grass and straw, and are often present in wood.

One xylan (that from esparto grass) comprises chains of about 19 xylose units with an arabinose residue at the ends, while in the xylan of wood hemicellulose there are 7 to 19 xylose units combined with a molecule of methylglucuronic acid. All these are classed as *pentosans.* Xylan is more readily attacked by bacteria than cellulose. It can be fermented by *"Aerobac." macerans* and *polymyxa,* yielding their typical fermentation products. It is also attacked oxidatively by three species of Cytophaga (Fuller and Norman, 1942). Probably many of the organisms active in retting can attack it.

A number of fungi and several bacilli excrete soluble enzymes that hydrolyse the water-soluble *pentosan* obtained from wheat flour (Simpson, 1954-1959). The presence of the pentosan itself, or wheat bran, xylan, etc., in the medium generally promoted formation of the enzyme, especially by *Bac. subtilis* and *pumilus.* Some sugars, especially xylose, and with *Asp. niger* arabinose, were also effective, so that stimulation of enzyme formation is by no means limited to substrates. *A gar* is a compound of rather peculiar structure, a polygalactoside containing galactose-6-sulphate. It is attacked slowly by only a few bacteria and fungi. Straw manure and especially marine algae often yield cultures whose colonies sink slowly into the agar, making a little well or depression. There is rarely any overall liquefaction like that of gelatin. Biochemically the attack can be shown by pouring iodine on the agar; the normal red-violet colour does not develop where the agar has been attacked. Colourless zones surrounding the colonies thus indicate formation of a diffusible enzyme, termed *gelase.*

The agar-liquefying organisms described are all aerobic and oxidise many sugars with little acid formation. Of seven species isolated from seaweeds, six attacked cellulose, but none attacked chitin (Stanier, 1941). The most active agar-liqueflers reported from soil are commonly vibrios, and this group includes one sulphate-reducer (Stuven, 1960).

The remainder are gram-negative rods (Gorseline, 1933; Buck and Cleverdon, 1960), an Actinomycete (Stanier, 1942), and several Cytophagas. Some of the Myxomycetes can liquefy agar too; this is particularly true for *Fullgo septica* and three species of Badhamia (Cohen, 1940).

The special carbohydrates forming the capsular polysaccharides of Pneumococcus are, of course, not immune to attack by other bacteria, and early in the study of the actions of bacteria on one another, Sickles and Shaw (1934, 1950) isolated a Bacillus *(Bac. palustris)* which attacks some of them.

The 2-N-acetyl-glucosamine of which it is a long-chain polymer occurs also in less polymerised form in the walls of bacteria. Because it occurs so widely, it is necessarily attacked by many organisms, including Aspergilli (Grassmann and Rubenbauer, 1931), some of the Basidiomycetes (Tracey, 1955), Streptomycetes, and several Eubacteria. These include Aerobacter, Serratia and Klebsiella. some Pseudomonads, Vibrios and two Clostridia (Clarke and Tracey, 1956), and the marine luminous *Photobacteria* (Spencer, 1961). Among the most active of all chitin-decomposers are Streptomycetes, which have been isolated from soil by enrichment cultures in media containing finely powdered chitin as sole organic constituent (Reynolds, 1954; Jeuniaux, 1951, 1958). These organisms, grown in a shaken aerobic medium, excrete a soluble stable chitinase, with optimum *pH,* 7, the amount of which is either dependent on, or increased by, chitin in the medium. Apparently it comprises three enzymes which act together in synergism (Jeuniaux, 1957). It does not remove the acetyl-group, but breaks down the chitin to N-acetyl- glucosamine or the corresponding biose, N,N-diacetyl-chitobiose (Reynolds, 1954). Formation of the biose parallels strikingly the production of maltose from starch (above).

Lignin is not a polysaccharide, but its breakdown may be mentioned here, since it is a major insoluble polymer of plant cell walls. It is an amorphous material of phenolic composition. The breakdown of lignin is not a hydrolysis but rather an oxidation and is most actively brought about in nature by white-rot fungi, especially Polyporus and Stereum species (Gottlieb and Pelczar, 1951). Although chemically processed lignins and their breakdown products have been found attackable by bacteria, native lignin is very resistant.

Virtanen (1946) found finely powdered wood to be attacked by some thermophiles in soil, and later Fischer (1953) obtained by enrichment cultures a group of gram-negative rods, probably Pseudomonads, forming red and brown pigments on potato, which attack lignin rather selectively; none could hydrolyse cellulose, and they oxidised phenol only weakly. Oxygen was essential, and addition of protein or peptone promoted the action. A number of Fusaria and other imperfect fungi were also active. There is some relation between the lignin bacteria and the humus-decomposing organisms.

Breakdown of Cellulose

Since it is the major insoluble and fibrous constituent of plant material, the rate of cellulose breakdown probably exerts a major influence on the carbon cycle in nature. Because cellulose also forms many of the commercial textile fibres, such as cotton, linen, jute, and hemp, its breakdown has been extensively studied, although in recent years emphasis has shifted from the bacteria to the fungi. The molecular weight of these fibre-molecules is from 200,000 to 1,000,000 (Pigman, 1957). It follows that there can be only a minute percentage of terminal reducing groups. This fact and its insolubility make cellulose resistant to many organisms. Soluble cellulases, producing hydrolysis to cellobiose, are formed in snails and a few fungi, but not in many bacteria.

The main active organisms are: Cytophagas and related aerobic Myxobacteria.

Certain aerobic sporeformers and non-sporeformers.

Clostridia.

The anaerobic non-sporeformers of the rumen of herbivores.

Several Actinomycetes.

Some of the Chytridiales, a primitive group of the Phycomycetes, that produce a soluble cellulase.

Many imperfect fungi, including (1) *Chaetomium globosum*, often used as a test organism in evaluating preservatives for textiles; (2) *Memnoniella echinata* and *Stachybotrys*, common on cotton fabrics; (3) *Myrothecium verrucaria*, one of the most active of the fungi, and common in soils, especially forest soils.

Detailed discussion of all these would be out of place, but four bacterial groups will be considered.

The Cytophagas

Although these organisms were found by van Iterson as early as 1904, in enrichment cultures with paper as the sole carbon source, their nature was not made clear till the detailed study of Hutchinson and Clayton (1919) at Rothamsted. These workers, from the knowledge that cellulose decomposition is fastest in a light, sandy soil, set out to look for aerobic cellulose-destroying organisms therein. Using as enrichment medium filter paper and salt solutions (with NH_4^+ as source of N), they obtained threadlike, slightly curved rods together with large micrococci, both being adherent to the paper fibres. The two forms could not be separated by heat, disinfection, or even spraying through an atomiser. When an almost pure culture of the thread forms was picked and diluted, the cocci eventually appeared even in the highest dilution. The organism therefore, exists in two forms, rod and sphere.

By placing filter paper on agar or silica gel containing the salts, and inoculating at intervals with tiny soil crumbs, Winogradsky (1929) obtained several such organisms distinguished by forming different pigments. However, in the form he studied in detail, only rods and no cocci were formed. The explanation for this, given by Krzemieniewska (1933), is that there are two types, those that (like Hutchinson and Clayton's organism) round up to form spheres (Sporocytophaga), and those that do not (Cytophaga proper). They all belong to the Myxobacteria, which are characterised by forming fruiting bodies through creeping together and piling up. The Cytophagas, however, are regarded as imperfect members in which fruiting-body formation is either reduced to the rounding up of single cells (Sporocytophaga) or is entirely absent (Cytophaga).

The Cytophagas were long believed to be limited to cellulose as a growth medium, but this was due to the toxicity of autoclaved sugar solutions. They will grow on glucose and cellobiose (when sterilised by Seitz-filter); *C. rubra,* a red form, also attacks xylose and mannose (Stanier, 1942a). The two marine forms attack agar and many sugars, but their action on cellulose is slow; they require peptone or yeast extract as nitrogen source, while the soil forms can use NH_4^+, nitrate, aspartic acid, or asparagine. Most Cytophagas are obligate aerobes. In cultures with paper or cotton dipping into a nutrient salt solution, the strongest attack is right at the liquid surface, where a strip of filter paper may be completely cut through in 7 to 10 days at 25° C. Only two species so far are facultative: *Cyt. fermantans,* which ferments glucose to acetate, propionate, and succinate, and *Cyt. succinicans,* which ferments glucose to pyruvate and, with the uptake of CO_2, to succinate (Bachmann, 1955; Anderson and Ordal, 1961).

Along with the Cytophagas, more complex Myxobacteria occur on cotton in the field (Heyn, 1957). Isolations on cellulose-salts agar showed that *Angiococcus cellulosum* and *Polyangium cellulosum* grow on raw cotton fibres. Both of these form large fruiting cysts, those of Angiococcus being yellowish pink and borne on stalks, those of Polyangium red to brown, stalkless, and up to 150 in diameter. The former, which is common, was first isolated from soil (Mishustin, 1938).

The mode of attack on cellulose is not understood. No reducing sugars are produced, and no cell-free cellulase has been obtained, although a preparation hydrolysing the polysaccharide, lichenin, has been reported (Greathouse, 1950). The organisms are closely adherent to the cellulose, the rods being aligned with the orientation of the microfibrils. If cotton preswollen in NaOH is used, and the cells washed off after growth, indentations can be seen where the cells have lain. Apparently, therefore, hydrolysis takes place at the surface of the cell, and the resulting sugars are consumed as fast as they are liberated. Polyangium penetrates the fibre so deeply that its fruiting bodies may be formed within the lumen (Heyn, 1957). The idea that cellulose destruction is simply a function of growth is supported by the observation that for every milligram of nitrogen taken up, about 30 mg of cellulose are destroyed (Hutchinson and Clayton, 1919).

Aerobic Eubacteria

The aerobes include the sporeforming Cellulobacillus, which yields a good instance of a bacterial cellulase. These organisms, of which two species, C. *myxogenes* and C. *mucosus,* were isolated, formed both glucose and cellobiose when grown in paper pulp in the presence of toluene (Simola, 1931). The cell-free culture fluid produced reducing sugar from cellulose and also hydrolysed cellobiose. As with Cytophaga, a good deal of the cellulose was assimilated by the bacteria; there were also formed coloured reducing acids related to humic acid, and, with uptake of nitrogen, a protein. As with Cytophaga, about 30 mg of cellulose were destroyed for every milligram of inorganic nitrogen taken up, which suggests that the basic process is similar in these otherwise unrelated types. There is also a group of non-sporeformers known as Cellulomonas; the name is unfortunate since some of them are peritrichously flagellated. These are probably the forms that were isolated from cellulose decomposing in sewage (Maki, 1954); they appeared to be the major cellulolytic organisms present. In pure culture they convert cellulose to alcohol, CO_2, and formic, acetic, and lactic acids. Their role in nature is not yet clear.

Anaerobic Sporeformers

Anaerobic cellulose breakdown has been studied ever since Omelianski (1902) obtained two sporeformers in cellulose-enrichment cultures from ditch mud. They had terminal spherical spores *(plectridia)*. Omelianski considered that the cellulose was fermented directly, without hydrolysis to sugars, but this was disproved when toluene or iodoform was added to the cultures, since the fermentation was then arrested, and both glucose and cellobiose could be identified in the medium (Pringsheim, 1912).

Pure cultures hydrolyse cellulose to glucose and cellobiose much faster than the sugars are subsequently fermented. Methane is often formed in enrichment cultures, doubtless due to a secondary fermentation. Pure cultures of these bacteria proved difficult to obtain. The insolubility of the cellulose makes dilution methods ineffective, and the requirement of the organisms for rich organic media often makes their growth on agar poor. It was not until 1944 that definitely pure cultures were obtained by Hungate. Thermophilic forms were later obtained by using partially degraded cellulose in colloidal or soluble form, incorporated in a rich agar medium and incubating at 55 to 60° C (McBee, 1950; Enebo, 1951, 1954). The cells produce a soluble cellulase, which yields mainly cellobiose and glucose; there is no soluble cellobiase, but most of the glucose is probably formed direct from cellulose (unlike the action of amylase).

The *Cl. thermocellulaseum* of Eriebo formed about the same products as Hungate's mesophilic form. Many years ago Khouvine studied the cellulolytic bacteria in the alimentary tract of man (1923) and various farm animals (1926).

By anaerobic enrichment cultures at 40°C, she isolated *"Bac. cellulosae dissolvens"* in what may have been pure culture. This organism, has somewhat unusual morphology: it is a long gram-negative rod with terminal oval "spores." It converts cellulose mainly to acetic acid and ethanol.

The products are similar to those of some clostridia from soil (Hungate, 1950). They are also similar to the products of some thermophilic forms which give 150 millimoles of acetic acid and 35 of ethanol from 100 "millimoles" of cellulose. Other clostridia gave a fermentation characterised by much succinic acid.

Two Clostridia, *Cl. lochheadii* and *Cl. longisporum,* isolated from the rumen of cattle being fed timothy hay, are extremely rapid cellulose digesters (Hungate, 1957). Like the above they form very long, apparently curved rods, but differ in producing large spindle-shaped clostridia in which the spores are formed. Again acetic acid and ethanol are prominent breakdown products of cellulose, along with CO_2, H_2, HCOOH, and in one case butyric acid. However, in general the sporeformers are not the main organisms responsible for cellulose utilisation in the alimentary tract of animals.

The thermophiles offer a striking contrast in rates. Omelianski's bacteria took three to four months at room temperature to ferment a few grams of cellulose; *Cl. thermocellum* began to produce gas from cellulose in 24 hours at 50° C, and a thermophile growing at 65° C (Woodman and Stewart, 1928) completed the whole fermentation in seven days. Most of these cultures were not pure.

Bacteria of Rumen of Herbivorous Animals

One of the most important uses of cellulose biologically is as feed for animals. Many of the herbivorous animals belong to the order *Ruminantia,* which means that the food after eating is transferred to a digestive pouch, the rumen, where it remains stirred and salivated for many hours before return to the mouth for chewing or passage on to the abomasum or "fourth stomach." In non-ruminants like the horse and pig, the caecum and colon are enlarged and doubtless play a similar part. Ruminants have very voluminous stomachs; the four organs together in cattle have a volume close to 250 litres, while that of the intestine is only 100 litres. Of the grasses and other plant foods, a third or more is cellulose, and a good deal of this is decomposed.

In the classic experiments of Thomas and Pringsheim (1918) a ram was fed for 11 days on potatoes and filter paper; determinations of paper fibre in the faeces showed that 57 per cent had disappeared.

Later experiments on steers showed a 90 per cent disappearance. Much of the carbon is converted to gas; a cow forms about 700 litres of CO_2 and methane daily. Since there are no cellulose-hydrolysing enzymes known in mammals, the decomposition is due to microorganisms acting in the rumen. Accordingly, numerous isolations have been attempted of bacteria living in the rumen of cattle and sheep. The first such studies always yielded enrichment cultures of Omelianski's organisms, rods with terminal spherical spores, producing acetic and butyric acids with CO_2 and hydrogen or methane. Khouvine's isolations from human faeces and the caecum of herbivores were mentioned above. Many others obtained similar organisms. In addition, an anaerobic actinomycete, *Micromonospora propionici,* has been isolated from the gut of termites (Hungate, 1946), where it decomposes cellulose. However, direct microscopic study of the flora of the rumen or caecum of herbivores did not support the importance of sporeformers in the breakdown of cellulose. Instead, the plant fragments mostly carried cocci of various sizes, often in chains like streptococci, together with vibrios (Baker and Martin, 1937-1939). These organisms gave a blue colour with iodine like that of the "granulose" of clostridia. The normal flora of the rumen and caecum was thus considered to consist in large part of coccoid bacteria (Baker, 1942), and approximate counts by the dilution method indicated that sporeformers were only present in very small numbers (Hungate, 1946).

The predominance of sporeformers in enrichment cultures is probably due to the change of conditions from those prevailing in vivo, and to the less exacting nutritional needs of the clostridia, although the two clostridia of Hungate (1957) probably do qualify as of real importance in cellulose breakdown in the rumen. By adding rumen fluid to the medium, cultures of a chain-forming coccus and a short ovoid gram-negative rod were obtained (Hungate, 1950; Sijpestijn, 1948, 1951). The great effectiveness of these in fermenting cellulose is shown by comparing rates; the sporeformers, even under their optimum conditions, took 13 days at 38° C for 80 per cent decomposition of added paper, while the cattle rumen cocci, in presence of rumen fluid, did the same in four days (Sijpestiin, 1948). Since four days is supposed to correspond to the actual time of digestion in cattle, the non-sporeformers evidently fit the physiological requirements. Sijpestijn's organisms comprised a gram-negative rod, *Ruminobacter parvum,* and a yellow gram-positive streptococcus, *Ruminococcus flavefaciens*. Hungate later (1959) obtained the gram-negative *Ruminococcus albus,* while *R. flavefaciens* was found in the caecum of rabbits (Hall, 1952). Often yeast extract could be substituted for rumen fluid. Two gram-negative rods of variable morphology isolated in a similar way were *Bacteroides succinogenes* (Hungate, 1950), which attacks cellulose, and *Bacteroides amylophilus* (Hamlin and Hungate, 1956), which attacks only starch and maltose.

Another form less active on cellulose appears to convert acetic to butyric acid and may be related to Butyribacterium. The difficulty of culturing Ruminococcus without rumen fluid is due to its nutritional requirements, which are like those of the lactobacilli; *R. flavefaciens* needs a complete amino acid mixture, folic acid, purines, cellulose or cellobiose, CO_2, and strict anaerobiosis (Ayers, 1958). Glucose is not used. With three of the five rumen forms listed, ethanol is less prominent than with the sporeformers.

It is perhaps just as well that the role of the sporeformers in the rumen is a minor one, since otherwise it might be difficult for cows to remain sober (up to 60 mole per cent yield of alcohol!). Three of the rumen bacteria apparently form either propionic or succinic acid as a second main product. The Actinomycete mentioned above forms propionic acid. Since Hungate's *B. succinogenes* (column 5) actually takes up CO_2, these two acids are in reality physiologically equivalent, by way of the decarboxylation of succinic acid. Succinate, indeed, is vigorously decarboxylated by washed rumen bacteria (Sijpestijn and Elsden, 1952) and also by pure cultures of the rumen micrococcus called *VeiHonella gazogenes* (Johns, 1951).

This organism, although probably the main source of propionic acid in the rumen, does not attack cellulose, or indeed any of the sugars. Presumably its main substrates are lactate and succinate formed by other bacteria. As to the other products, it is evident that they vary with the organism. Some of the bacteria (especially those in column 7) produce reducing sugar in moderate amounts, but this has been largely corrected for. Small amounts of aromatic fatty acids are found, these being produced by deamination of tyrosine and phenylalanine. The products are altered, too, by differences in the CO_2 tension, which influences the decarboxylation of both succinic and pyruvic acids. Low CO_2 tensions would thus favour the formation of alcohol, high tensions that of succinate.

It will be noted that methane is not a product in these pure culture fermentations. On the other hand, methane does appear in fermentations carried out by the total bacteria centrifuged out of rumen fluid (Marston, 1948). As far as is known, the methane fermentation is undergone only by organic acids or alcohol, and not by cellulose. The methane that is produced in the rumen, therefore, is due to a secondary fermentation of the products, and two Methaiiobacteria have been isolated from rumen that do this: *Mb. formicicurn*, which utilises formate (Opperman *et al*. 1957), and *Mb. ruminantiuni* which is present in larger numbers, up to 200 million per millilitre, and makes methane most actively from hydrogen and CO_2 (Smith arid Hungate, 1958).

These are very strict anaerobes, especially the latter, which requires dithionite, or growing *E. coli*, to make the redox potential sufficiently negative for growth. The conclusion that the rumen contains many organisms, closely related physiologically, and mostly of short rod, oval, or coccus morphology, which actively ferment cellulose to organic acids, agrees well with studies of cattle nutrition, which have shown the presence of large amounts of acetic and propionic acids in the rumen after cellulose ingestion. Both these acids can be converted in the animal body into glucose, glycogen, and (in the case of acetic acid) to fat. Further, the bodies of the bacteria, rich in granulose, should also be assimilable after hydrolysis. These comprise from 7 to 18 per cent of the total combustible energy of the cellulose.

It has been calculated from fermentations with combined rumen bacteria that at least 70 per cent of the energy requirement of cattle comes from the fermentation acids of the rumen (Carroll and Hungate, 1954) and cellulose provides a good part of these. It is clear, therefore, that ruminal and intestinal cellulose breakdown is an important and effective part of the natural carbon cycle.

4

Nitrogen Cycle

Nitrification and Denitrification

Nitrification Process and its Causative Organisms: The end product of protein breakdown, so far as the nitrogen is concerned, is ammonia. Yet ammonia does not accumulate in the soil, since most of the soil nitrogen is in the form of nitrate. Since solutions of ammonium salts do not change to nitrate when simply exposed to the air, it was early realised that ammonia must be converted to nitrate only under very special conditions. Solution to the problem came through the finding that the process is biological, not chemical.

Schloesing and Miintz (1877-1879) in France passed sewage water very slowly through a column of sandy soil and chalk, and found that after 20 days ammonia no longer issued at the other end, but its place was taken by nitrate.

The process ran for four months. The reaction was very slow at 50° C and had its optimum temperature at 30° to 35°C. It was stopped completely by drawing through the tube either chloroform or boiling water.

However, when, after boiling, a little fresh soil was added to the column, the power of oxidising ammonia soon returned. They deduced, therefore, that the conversion of ammonia to nitrate, or as it is usually called *nitrification,* is due to living creatures. In parallel studies with solution cultures, Warington (1878-1891) confirmed the French workers' findings and established the following facts:

1. A minute seeding (0.1 gm of soil in a large flask of sterile ammonium sulphate solution) suffices to start nitrification.
2. Seedings of soil taken from more than 18 in. in depth were inactive, i.e., the organisms must be only in the upper layers of the soil.
3. The ammonium sulphate solution without seeding keeps indefinitely.
4. Small amounts of phosphate are needed; this recalls the slogan based on the analysis of animal tissues, *"Ohne Phosphor kein Leben."*
5. The solution becomes acid during the process, because the NH_4^+ cation is being converted to the ~NO_3 anion and hydrogen ions. Addition of $CaCO_3$, therefore, promotes nitrification by providing a reserve of base. (Its second function, that of providing essential Ca ions, was not discovered until 50 years later.)
6. Ordinary bacteria isolated from air do not nitrify, so that a special organism must be concerned. Schloesing and Miintz also had no success using highly oxidative fungi.
7. Organic materials such as urine and milk could only be nitrified when in high dilution, and then always more slowly than ammonium salts. Addition of glucose also slowed the reaction.
8. Sometimes the reaction went only as far as nitrite. Increased aeration did not affect this. Successive seedings from the nitrite-producing solutions gave only

nitrite, but when fresh soil was added, nitrate quickly appeared.

9. The complementary phenomenon of seedings which only converted nitrite to nitrate and did not oxidise ammonia was also occasionally found.

It followed from this work that two kinds of special, aerobic organisms must be responsible, respectively, for the two stages of oxidation, $NH_3 \rightarrow HNO_2$ and $HNO_2 \rightarrow HNO_3$. However, all attempts to isolate these organisms from the enrichment cultures failed.

The method which had just been introduced by Koch, of plating out in nutrient gelatin medium, although brilliantly successful with other organisms, gave no satisfactory results here. Several other workers were equally unsuccessful. Although numbers of very short rods could be observed in the nitrifying solutions, no colonies could be obtained on nutrient gelatin.

The problem was finally solved through the introduction of a quite new viewpoint by the young Russian, Winogradsky (1890, 1891). The difficulties of the problem, and the originality and resourcefulness brought to bear upon it, have made his researches a classic of microbiology. Pointing out that the oxidation of ammonia yields energy, Winogradsky realised that this energy could be applied to the formation of organic matter by *reduction of* CO_2. The process would resemble photosynthesis, but, instead of light, the energy required would be supplied by a chemical oxidation.

This idea was supported by the inhibiting effect of organic matter (noted by Warington) and by Winogradsky's finding that nitrification proceeded most vigorously in the absence of any added source of carbon except carbonate. The conception of the nitrifying bacteria as *chemosynthetic*, i.e., synthesising their organic matter through energy supplied by a chemical reaction, at once enabled their isolation.

Using a gelatin medium, free from organic nutrients and

containing only $(NH_4)_2SO_4$ and other inorganic salts, Winogradsky made inoculations from the small floccules attached to the $CaCO_3$ particles in liquid enrichments. From these, colonies developed which were not the nitrifying organisms. By then picking from where no visible colonies grew and inoculating into fresh inorganic medium, he obtained pure cultures of the nitrifiers. He afterwards found that they would form colonies, although only minute ones, if the medium were solidified with silica gel.

Two types were isolated, *Nitrosomonas,* oxidising ammonia to nitrite, and *Nitrobacter* oxidising nitrite to nitrate. Examination of soils from all over the world led to the recognition of two species of Nitrosomonas, *N, europeaensis,* from Russian, Swiss, and French soils, and TV. *javenensis,* from East Indies soil. The various strains of Nitrobacter were all indistinguishable. As the photographs show, Nitrosomonas is a very short rod with a long polar flagellum. In a later return to the subject, Winogradsky and his sister (1933) distinguished four Nitrosomonas species, a *Nitrosococcus* and a spirillum, *Nitrosospira.*

From pine-forest soils, on which nitrification is very slow (largely due to the acidity), some organisms forming cystlike zoogloeae, *Nitrosocystis, Nitrosogloea,* and *Nitrocystis,* were isolated and V. *javenensis* placed in the first of these genera. These organisms are probably Myxobacteria and not nitrifiers at all, because Imsenecki (1946) found that nitrifying cultures were readily overgrown by a member of the genus Sorangium forming hard, angular fruiting bodies. Bisset and Grace (1954), therefore, assign both these "genera" to Sorangium and consider them non-nitrifiers.

Others believe them to be "hard-colony" variants of Nitrosomonas. When it is considered, too, that Actinoplanes forms sporangia from which polarly motile "spores' emerge, one wonders how many possible organisms may be involved. All in all, Nitrosomonas and Nitrobacter are the only well-defined and active nitrifiers. The isolation and pure culture of nitrifying bacteria, although often repeated are tedious by ordinary methods.

Even after continued transfers in inorganic media from the enrichment cultures, Nelson found that nitrifiers were still in a minority of 1 to 1000. What the other 999 organisms were living on, in a purely inorganic medium, is not clear, but it seems certain that the slow growth of the nitrifiers is due to inadequate aeration (Lees, 1951). Since their oxygen requirement for growth is so much greater than that of ordinary bacteria (up to 100 atoms of oxygen per atom of carbon assimilated), their growth is greatly favoured by growing them on glass beads percolated at intervals with the liquid medium.

This ensures a very thin layer of medium and free access to air. While the simple process of plating and picking colonies from flask cultures would scarcely ever give pure cultures, the percolation method, after only five transfers, gave cultures apparently free from heterotrophic organisms.

The difficulties of culture are increased by the fact that the colonies are minute and are firmly attached to the silica gel. They can, however, be enlarged in size by adding traces of peptone (Fred and Davenport, 1921), *"Nahrstoff Heyden"* a peptone product (Kingma Boltjes, 1935), or even soil extract (Hes, 1937). The rate of nitrification, however, is not enhanced by soil extract or peptone, and the latter, even at 0.2 per cent, is quite toxic (Meiklejohn, 1952). Two older methods are worth mention. In one, silica gel plates, soaked in a salt solution containing ammonium sulphate, are coated with a cream of $CaCO_3$, which is allowed to dry (Winogradsky, 1933).

Minute grains of soil are then placed at intervals on the plate. After 6 to 15 days, clear zones appear around some grains, owing to the nitrous acid, which dissolves the $CaCO_3$. By picking from these zones, making serial dilutions in water, and again plating, pure cultures can be obtained. In addition, by counting the percentage of grains that develop clear ones, the numbers of nitrifiers in different soils can be compared.

In the other method (Engel and Skallau, 1937) a very large number of transfers (say 100 or more) are made from one vigorously nitrifying liquid culture; on individual examination, one or more of these may be found to be pure cultures by their ability to nitrify but inability to produce any growth in broth. Probably neither of these can compete with Lees' percolation procedure.

Influence of pH and of Substrate

Meyerhof (1916, 1917) found by measuring oxidation rates that the optimum *pH.* of Nitrosomonas is 8.5 to 8.8, and that of Nitrobacter about the same. The steep decrease in rate on the acid side has not been wholly confirmed, because Hofman and Lees (1953) found a much more gradual fall, reaching 50 per cent of the maximum rate at *pH,* 6.5. For Nitrobacter they also found the curve somewhat more to the acid side, the optimum *pH.* being 7.7. The fall in rate on the alkaline side may be due to the inability of the cells to use the "CO_3 ion as source of CO_2, while they could use HCO_3".

The narrow pH optimum, with a steeper fall on the alkaline than on the acid side, favours this view. The suggestion that the fall on the alkaline side is due to penetration of Nitrosomonas cells by free NH_3 (Meyerhof, 1917) can hardly be valid, because Nitrobacter shows a similar sensitivity to alkali.

Indeed, in the case of Nitrobacter, ions must enter the cell freely, since: (1) at the optimum *pH* both nitrite and bicarbonate are very largely dissociated, and (2) unionised NO_2 and NO groups in the form of metallic complexes or organic nitroso-compounds (Meyerhof, 1917) could not be used.

The original medium of Winogradsky (1890, 1891) used $MgCO_3$ instead of $CaCO_3$. When this was made up with purified chemicals, it gave very poor growth and nitrification. In an elegant study of the influence of impurities, Kingma Boltjes (1935) established that the missing factor was $CaCl_2$ Apparently the Ca ion is essential for Nitrosomonas. Solid $CaCO_3$ thus fulfils two functions, that of control and of Ca supply. Solid $CaCO_3$ is not necessary for Nitrobacter, however (Meyerhof, 1916).

Part of the difficulty in culturing nitrifying bacteria doubtless resides in their sensitivity to excess substrate, when in cultures. Nitrobacter is inhibited by nitrite and Nitrosomonas by ammonia. The inhibition may be slowly overcome, but is powerful at first. The oft-quoted data of Meyerhof have been recalculated as absolute concentrations and plotted on a logarithmic scale; the sharp substrate optimum for both bacteria is remarkable and is quite unlike the behaviour of most other organisms with their substrates.

What happens is that the efficiency of Nitrosomonas decreases as nitrite accumulates. Correspondingly, growth of Nitrobacter slows down as nitrate accumulates, but if the nitrate is continually dialysed away, as much as 6 gm nitrate per litre can be formed without loss of activity (Gould and Lees, 1959). Nitrobacter is also inhibited by free NH_3 (Meyerhof, 1916), and, therefore, it will often not develop until Nitrosomonas has lowered the NH_3 level. When most of the NH_3 has disappeared, nitrate suddenly begins to be formed. Lees and Simpson (1957) have to some extent circumvented these inhibitions by using an aerated suspension of metal phosphates as medium. The cells remain attached to the particles, which can be easily washed; the suspensions are convenient material for experiments. Both forms are inhibited by divalent metal ions, the inhibition of Nitrobacter increasing sharply with *pFL* (Meyerhof, 1916, 1917), and also by lowered oxygen pressure (Meyerhof, 1916; Amer and Barthọlomew, 1951), which suggests that at least one stage of the oxidation involves an oxidase with much less affinity for O_2 than cytochrome oxidase.

Influence of Organic Matter

Since the nitrifiers are autotrophic, it is easy to see why their growth is not supported by organic matter. But what is remarkable is that many organic compounds actually inhibit nitrification. Not only gelatin, but many amines, alkaloids, and even amino acids inhibit. Guanidine and aniline derivatives were particularly potent, both in solution cultures (Meyerhof, 1917) and in soil enrichments (Lees and Quastel, 1946).

Metal-combining agents such as cyanide, thiourea (Quastel and Scholefield, 1951), and especially allylthiourea (Hofman and Lees, 1953) inhibit, as also do the urethanes. Hydroxylamine, a possible intermediate, also inhibits somewhat (Meyerhof), although it is itself slowly oxidised. Early workers found sugars inhibitory too, but this has not been confirmed, and the inhibition caused by glucose is in fact due to a decomposition product formed on autoclaving (Jensen, 1950). Thus, it appears that the strongest inhibitions are due either to interfering with the metal-enzyme oxidase system (cyanide, etc.) or to chemical relationship with the substrate ammonia (the organic bases).

Antibiotics seem not to affect nitrification directly. Concentrations of streptomycin that strongly inhibit growth have no direct effect on ammonia oxidation (Lees, 1952). The actual growth promotion by Nahrstoff-Heyden is in marked contrast to the general inhibition by nitrogenous compounds. Nitrobacter is less sensitive to organic compounds, and its colonies can be obtained on nitrite agar; they are very small, tough, and difficult to transfer. Besides excess nitrite, it is inhibited by cyanide and azide (Lees, 1954), by chlorate (Meiklejohn, 1952), and among organic compounds, by methionine, nitrourea, thiourea, and antibiotics (Quastel and Scholefield, 1951).

Nitrification in Soil

Nitrification usually proceeds fast enough that little free ammonia remains in soil, although some grass-clover pastures have been found to have about equal amounts of ammonia and nitrate (Richardson, 1938). Correspondingly, the nitrate is reduced back to NH_3 and amino acids fast enough in the plants that they rarely contain more than traces of nitrate, which is fortunate because nitrate and nitrite are toxic to animals.

After a long dry summer has rendered the soil very aerobic, rains may cause so much nitrification that the crops become rich in nitrate; this seems to occur often in New Zealand (Bathurst and Mitchell, 1958; Butler, 1959) where rye-grass (*Lolium* spp.) and oats reach high nitrate levels.

The fluctuations in nitrification rate and in soil nitrate are evidently very wide. Neither in pure nor in enrichment cultures can any of the nitrifiers yet known grow at *pH* more acid than 6. In the extensive Rothamsted trials with grassland, soils more acid than 6 tended to accumulate NH_3 (Richardson, 1938). This raises the question of how nitrification can proceed in acid soils. It does take place, though slowly. No doubt the nitrifying bacteria act mainly when attached to mineral particles, i.e., at points where the pH is locally less acid.

This is easily visualised from the discussion which shows how heterogeneous the conditions of growth in soil usually are. An alternative explanation was advanced by Indian workers who claim that, in tropical soils at least, ammonia may be oxidised photochemically (Dhar and Rao, 1933).

Zinc oxide and TiO_2, which are present in many soils, were claimed to act as catalysts. However, in view of the stability of ammonium salts in the presence of sterile soil, it is unlikely that this factor could be of general importance.

The attachment of nitrifying bacteria to particles explains also the nitrification in soil in spite of the presence of organic matter. For on a soil particle, local areas can be freed of organic matter by oxidising organisms and thus made suitable for the growth of nitrifiers.

The same holds for the sand filters of sewage outfall works, which must be well supplied with organic matter, but from which Nitrosomonas can be readily obtained. In this connection it is worth noting that marine nitrifying bacteria (i.e., salt-tolerant races) can be obtained from the sea bottom, but very seldom from sea water itself (Issatschenko, 1926; Carey, 1938). For several years Quastel and co-workers studied nitrification in soil under laboratory conditions by continually perfusing solutions through a small column of soil on a filter.

The technique is reminiscent of that of Schloesing and Miintz, but aeration is much more efficient, and the apparatus is cyclic and completely enclosed. The soil perfused with ammonium salt becomes an enrichment culture, and the rate of nitrification increases steadily during the perfusion. If the process is stopped and then restarted with fresh solution, the rate begins at once at the maximal value previously reached and does not increase further (Lees and Quastel, 1946). This means that the surfaces of the soil particles are now saturated with the organisms.

With these enriched soils many of the findings on liquid cultures have been readily reproduced. The most important findings (Quastel and Scholefield, 1951) are:

1. Calcium carbonate improves nitrification on most soils. Sodium salts of organic acids have a similar effect owing to their rapid conversion to bicarbonate.
2. Although amino acids in general do not appreciably inhibit, methionine and methionine sulphoxide do.
3. Many inhibitors exert quite different action on the two processes. Urethanes, guanidines, thiourea, and p-aminosalicylic acid inhibit ammonia oxidation; nitrourea, thiourea, and chlorates inhibit nitrite oxidation, but urethanes do not. Since chlorate does not affect Nitrosomonas, soils in presence of N/1000 chlorate accumulate nitrite, and this permits study of ammonia oxidation alone.
4. Enrichment of Nitrosomonas and Nitrobacter on perfusion with nitrite takes place only after a lag, which is doubtless due to the inhibitory effect of nitrite on the organisms. That they develop at all indicates either selection of a nitrite-resistant strain or development of a protective mechanism.
5. Amines and amino acids in general are only nitrified

after conversion to ammonium. An exception, however, is pyruvic oxide, CH_3C-COOH, which appears to be converted directly to nitrite by the NOH soil enrichments. This conversion is not due to Nitrosomonas, however, but to three groups of heterotrophic bacteria (Jensen, 1951). A proactinomycete, *Nocardia corallina*, is the most active of these.

Nitrogen: Carbon Ratio and the Biochemistry of Nitrification

By comparison with other bacteria, the nitrifiers are not highly active. A Nitrosomonas cell takes nearly an hour to produce twice its weight of nitrite (Engel, 1929). Yet Urobacillus hydrolyses 1000 times its weight of urea in an hour. On the other hand, since Nitrosomonas can live only by an oxidative process, the oxygen consumption is relatively high. As a result the yield of cells per unit of time, or per unit of oxygen consumed, is low.

Autotrophic growth with ammonia oxidation can be considered as a pair of linked reactions, the second of which has two alternatives:

a. Oxidation by oxygen: $12(H) + 3\ O_2 \rightarrow 6\ H_2O$

b. "Oxidation" by CO_2: $12(H) + 3\ CO_2 \rightarrow 3(CH_2O) + 3H_2O$

Winogradsky measured the ratio N:C (i.e., NH_4+ oxidised:CO_2 reduced) and found it to be about 35. By similar methods, Nelson (1931) obtained 15; Hes (1937), values close to 33; and Bomeke (1951), 35 to 53. Old cultures give higher figures, i.e., they are less efficient in assimilating carbon (Hes, Bomeke). Apparently there is a steady increase in the figure with increase of nitrite in the solution. Hofman and Lees (1952) have found that the results of different workers, with different amounts of nitrite formed, all agree with the approximate equation:

$$\log C = 0.73 \log N - 0.35$$

where C = organic carbon formed and N = nitrite formed, both in milligrams per litre. On thermodynamic grounds, but assuming rather different (and improbable) reactions, Baas-Becking and Parks (1927) calculated that with a ratio N:C of about 30, Nitrosomonas is only about 5.9 per cent efficient. That is, of the energy liberated by ammonia oxidation only 5.9 per cent is used for CO.

However, in *E. coli* the galactosidase can be induced not only by lactose but also by the reduction. For Nitrobacter they found about 9 per cent.

However, it follows from the equation that very young Nitrosomonas cultures would be much more efficient. Actually, when the nitrite is below 20 mg per litre, efficiencies up to almost 50 per cent have been found (Hofman and Lees, 1952). The free energies are such that the organism should need to oxidise only 2 or 3 molecules of ammonia to reduce one of CO.

The wide variation and the influence of age and nitrite suggest that the method of linkage between reactions is a highly sensitive one.

It has been suggested that the low thermodynamic efficiency is due to loss of organic matter by respiration, but the respiratory rate of the nitrifiers is extremely slow (Bomeke, 1939); Nitrobacter oxidises formate one-fifth as fast as nitrite and also (more slowly) acetate (Silver, 1960).

It is more likely that whatever high-energy linkage is produced as a result of the oxidation of ammonia is very easily destroyed or sidetracked before it has caused reduction of CO_2. Perhaps some of the strong inhibitions by organic compounds are due to their reacting with this linkage.

Another sign of sensitivity is that resting cultures of Nitrobacter soon lose their ability to oxidise nitrite; on supplying nitrite, O_3, and CO_2 (all three essential), the power returns, but only after 40 hours or so (Seeler and Engel, 1959).

The inhibition of respiration by pure O_2, observed by Meyerhof, suggests peroxide accumulation and therefore participation of a flavoprotein enzyme. As to the first, thiourea and allylthiourea inhibit the oxidation of ammonia but not that of hydroxylamine (Hofman and Lees, 1953). Added hydroxylamine is rapidly oxidised and without any time lag (Engel and Alexander, 1958). Hydrazine, on the other hand, prevents the formation of nitrite, and instead some hydroxylamine accumulates.

Nitrate Reduction or Denitrification

Flora of Nitrate Enrichment Cultures: When mixed fermentations go on in the presence of nitrates, it is common for nitrite and even N_2O or N_2 to be produced.

This was observed as long ago as 1868 by Schloesing and was first studied using a mixture of KNO_3 with ammonium citrate and asparagine (Gayon and Dupetit, 1882).

Two organisms were isolated, one (probably *Ps. pyocyaned)* producing N_2 and N_2O, the other *(Ps. stutzeri)* forming mainly N_2. In essence, the reduction of nitrate is the use of the oxygen of nitrate as a hydrogen acceptor.

Consequently, a source of combined hydrogen and a limitation on the supply of free oxygen are needed. Recognition of these two conditions by Beijerinck and Minkman (1909) led to their making enrichment cultures from soil under anaerobic conditions but in the presence of KNO_3, using various organic compounds as hydrogen donors.

They found, however, that inclusion of a small bubble of oxygen at the start ("stimulus oxygen") improved the growth. Much nitrous oxide was evolved. A crude culture (mostly sporeformers) decomposed in three days the whole of 2 gm KNO_5 added, producing 159 ml N_2O, 75 ml N_2, and 11.5 ml CO_2. Since the theoretical yield of N_2 or N_2O is 222 ml, the agreement is within 5 per cent.

Both the composition of the gas and the bacterial flora obtained depend on the nitrate concentration and the carbon source. In broth containing 5 to 12 per cent KNO_3 (0.5 to 1.2 M), with alcohol or organic acids as hydrogen donor, the gas contains much N_2O, and sporeformers (such as Beijerinck's *Bac. nitroxus)* predominate. When protein is used as hydrogen donor, e.g., in a study of denitrifiers growing in cans of salted ham (which contains KNO_3), thermophilic sporeformers resembling *Bac. subtilis* were obtained (Verhoeven, 1950). All of these are now regarded as strains of one species, *Denitrobacillus licheniformis* (Verhoeven, 1952).

In media containing only 1 to 2 per cent KNO_3, mostly Pseudomonads develop. With alcohol or propionate as hydrogen donor, *Ps. aeruginosa* often appears, giving a gas containing about two-thirds N_2O; with other organic acid salts, *Ps. stutzeri,* the gas from which is largely nitrogen, and *Ps. vulpinus* are obtained (van Iterson, 1903).

Tartrate gives rise to *Ps. stutzeri* with great regularity, and pure cultures can be obtained in two transfers (van Niel and Allen, 1952).

With sugar as hydrogen donor, Korsakowa (1927-1929) obtained a strain of *Ps. fluorescens,* giving almost wholly nitrogen. Even hydrogen gas can be used as donor in the special case of M. *denitrificans* (Kluyver, 1953).

From sea water an organism giving almost pure N_2 has been isolated (Lloyd and Cranston, 1930); as with Beijerinck's cultures, a trace of oxygen was necessary to start growth, but nitrate reduction did not take place in aerobic conditions.

Among other denitrifiers the Rhizobia should be mentioned (Wilson, 1947), while very many facultative anaerobes, especially the coliform bacteria, can reduce nitrate to the stage of nitrite.

Physiology of Nitrate Reduction

1. *Competition With O_2:* Since, the reduction of nitrate (as also of nitrite) is essentially the use of bound oxygen as hydrogen acceptor, it might be expected that free oxygen would inhibit nitrate reduction. Experiments

of this sort with *E. coli,* using the reduction only to nitrite, gave the data shown in table below:

Table: Effect of O_2 on Nitrate Reduction by *E. coli*

Oxygen in Gas (per cent)	*Inhibition of Nitrate Reduction (per cent)*
0	0
0.4	21
1.1	61
3.8	93
21	94
99	96

The standard test of the SAB for nitrate reduction in more or less aerobic broth cultures is therefore almost meaningless, since the extent of nitrite formation will depend on whether the growth is strong enough to consume the dissolved oxygen. The phenomena of selection and of induction also complicate the test.

2. *The Nature of the Enzyme System*: In order for the oxygen of nitrate to combine with the hydrogen of organic compounds, it has to be activated by a special enzyme, called nitrate reductase or *nitratase.* Evidence for this comes from inhibition experiments. Cyanide strongly inhibits reduction of both nitrate (Stickland, 1931) and nitrite (Elema et al., 1934; Yamagata, 1939), and so does carbon monoxide. In *E.* coli, nitrite reduction is inhibited also by iodoacetate (dehydrogenase type of poisoning), so that it behaves just as normal oxidation with O_2 would (Aubel, 1938). Evidently, therefore, the process requires, as well as a dehydrogenase, a metalloenzyme of the oxidase type. This enzyme is a complex of the nitrate-reducing system and the cytochromes. Nitrate-reducing bacteria are rich in cytochromes; *Ps. stutzeri* growing on potato forms so much cytochrome that the

normally colourless colonies become pink (van Niel and Allen, 1952); so do those of *Ps. aeruginosa* grown anaerobically with nitrate (Verhoeven and Takeda, 1956).

The formation, both of cytochromes and of nitratase is readily inducible by the medium. Nitratase appears in *E. coli* when the cells are incubated in mannitol and nitrate; no growth need occur, but amino acids and Mg are necessary for formation of the enzyme (Pollock, 1946; Pollock and Wainwright, 1948), indicating that protein synthesis is involved. When mutants requiring a specific amino acid were used, that amino acid was also needed for the formation of nitratase (Wainwright and Nevil, 1956). *Ps. aeruginosa* grown on nitrate is rich in nitrate-reducing enzyme, while on nitrite it forms much of the nitrite-reducing system that yields N_2. This last observation shows that there are two enzyme systems, *nitratase* and *nitritase*. The first is readily isolated by suspending *E. coli* in fluoride-phosphate mixture, when the enzyme goes into solution (Yamagata, 1939), or by breaking up the cells (Nicholas and Nason, 1955). After it was shown that nitratase from green leaves and from Neurospora contains molybdenum and flavoprotein, similar properties were shown for the *E. coli* enzyme, and cells grown on a medium free of Mo were found deficient in nitratase (Nicholas and Nason, 1955).

The cell-free enzyme uses DPNH to reduce nitrate. When isotopic Mo or Fe was present in the medium, the metal became concentrated in the purified enzyme (Fewson and Nicholas, 1961). Furthermore, when repeated precipitation had lowered the activity, it was restored by adding flavin-adenine-dinucleotide, FAD. The enzyme contained cytochrome c, which was seen to be reduced by the DPNH and reoxidised by nitrate or molybdate. It can also be reoxidised with N_2O itself (Verhoeven and Takeda, 1956). Correspondingly, if the oxidase is inhibited by CO, aeration no longer prevents nitrate reduction; i.e., nitrate (via the Mo) and oxygen (via the oxidase) *compete* for the reoxidation of the reduced Cytochrome *c*. *Nitritase* appears to contain no Mo, but contains a flavin (Taniguchi *et al.*, 1952), and that from *Ps. stutzeri* contains both Cu and Fe (Chung and Najjar, 1956). While the cells of *Ps. aeruginosa* convert nitrate to No, the cell-free enzyme after purification forms only NO (Walker and Nicholas, 1960); that from *Ps. stutzeri* forms N_2 and NO (Chung and Najjar, 1956). Using reduced FAD as hydrogen donor the reaction is:

$$FADH_2 + 2\ NO_2 \longrightarrow 2\ NO + FAD + 2\ OH$$

The preparation of Chung and Najjar also reduced NO, using DPNH or TPNH and a flavoprotein. Although the enzymes evidently contain a Cytochrome c, it did not appear to be reduced in the process.

3. *Substrates:* If nitratase operates as an alternative to cytochrome oxidase, any substrate oxidisable through the Cytochrome system would be expected also to be oxidisable with nitrate reduction. Alcohols, organic acids, and sugars can be so used, depending on the organism and the nitrate concentration. In addition, many strains of *E. coli* use H_2 for the reduction to nitrite (McNall and Atkinson, 1956), and the case of *M. denitrificans,* for the full reduction to N_2. Glucose as substrate is completely dehydrogenated by *Ps. aeruginosa* to CO_2 and H_2O, the nitrate being converted to N_2 with about 10 per cent of N_2O (Verhoeven and Goos, 1954). In one rare case, even sulphur reacts. Two groups of compounds are exceptions to this rule: aromatic acids and aliphatic hydrocarbons. The reason for the first is clear; oxidation of the benzene ring requires the introduction of OH groups, and therefore *oxygen atoms* are required and are incorporated. To provide oxygen is not the same as to accept hydrogen.

As to the second case, *Ps. stutzeri* will oxidise several hydrocarbons in air, but not anaerobically with nitrate (Hansen and Kallio, 1957); however, their corresponding alcohols *can* be oxidised with nitrate. Hansen and Kallio, therefore, believe that the first step in hydrocarbon oxidation is the introduction of OH groups. Evidently, then, nitrate can act as *hydrogen acceptor* but not as *oxygen donor.*

4. *The Intermediate Stages:* Three stages could be recognised:

$$HNO_3 + 2(H) \longrightarrow HNO_2 + H_2O$$

$$2HNO_2 + 4(H) \longrightarrow (H_2N_2O_2) + 2H_2O$$

$$(H_2N_2O_2) + 2(H) \longrightarrow N_2 + 2H_2O$$

The postulated intermediate product in equations (2) and (3) is unstable and could alternatively break down thus:

$$(H_2N_2O_2) \rightarrow H_2O + N_2O$$

This is, no doubt, the source of the nitrous oxide so often produced, especially by the sporeformers. In general, high temperatures lead to N_2O production; at lower temperatures the $(H_2N_2O_2)$ presumably survives long enough to be further reduced. For example, Denitrobacillus produced 49 per cent N_2O (with 10 per cent CO_2 + 41 per cent N_2) at 30° C, rising to 63 per cent N_2O at 51° C (Verhoeven, 1950).

Low pressures also favour N_2O formation, by allowing the gas to escape as fast as it is formed; high pressures correspondingly prevent N_2O formation, and nitrogen appears instead (Verhoeven, 1952). The denitrification process has been followed in anaerobic cultures by oxidation-reduction measurements (Elema, 1932).

Since anions are being removed, the *pH.* becomes alkaline, so that the potentials must be compared with those of a hydrogen electrode in the same solution. As soon as nitrite appears, the potential rises to about 475 mv above the hydrogen electrode. When all nitrate is reduced, the value drops about 25 mv and remains there until all nitrite is used up; then it falls precipitously. If at the end more nitrate is added, the potential rises again to the nitrite level, to fall again when all nitrite is consumed (cf. Elema *et al.*, 1934).

It may be concluded that there are two characteristic potentials (i.e., reducing tendencies); one for the nitrate-nitrite system and a lower one for the nitrite-$H_2N_2O_2$ system. In these experiments the necessary reversible oxido-reduction system to react at the electrode was furnished by some excretion product of the bacteria *(M. denitrificans* or *Ps. vulpinus)* so that addition of a reversible dye was not necessary.

These experiments show that the postulated intermediate between nitrite and oxygen really exists, but they give no evidence as to what it is. It cannot be N_2O, because *Ps. denitrificans*, which reduces nitrate rapidly to nitrogen without forming any appreciable N_2O, is unable to reduce N_2O itself until after a time lag (Sacks and Barker, 1952). This indicates that it has to form a special induced enzyme for N_2O reduction; the formation of this *enzyme*, like that of many other induced enzymes, is inhibited by azide or dinitrophenol. Yet these reagents do not inhibit the conversion of nitrite to N_2. Besides, in *Ps. stutzeri* the different reductions show different sensitivity to cyanide.

For example, M/100 KCN at *pR* 7.3 completely inhibits the reduction of N_2O, but only causes 60 per cent inhibition of the reduction of nitrite (van Niel and Allen, 1952). In other words, nitrite can be converted to nitrogen under conditions where N_2O cannot. Alternative intermediates are nitroxyl, NOH, hyponitrous acid, HONrNOH, nitramide, NH_2NO_2, and nitric oxide, NO. Evidence for one or two of these has been brought up from time to time and then discarded. Hyponitrite is attacked by a few denitrifiers, but *Ps. stutzeri*, one of the most active, produces no more gas from it than is yielded by spontaneous decomposition (van Niel and Allen, 1952). The fact that NO is formed by two of the cell-free systems, and can be reduced also by them, makes it a strong candidate for the position of intermediate, although like N_2O it could lie just off the main pathway. However, in the case of *Thiobacillus denitrificans,* NO is produced best in presence of M/1000 KCN, and in the case of *Ps. aeruginosa,* by cells more than 40 hours old (Verhoeven, 1956). Furthermore, NO is reduced very slowly and not only by denitrifiers but also by *Ps. fluorescens* and even yeast. For these reasons NO is probably not an intermediate either, and it seems that the true intermediate remains bound to the enzyme. The case is oddly parallel to that of the reduction of nitrogen itself.

The production of alkali in the process is important. Emphasis has been laid on denitrification in the sea as a cause of calcium carbonate precipitation, because, if sea water (containing Ca $[HCO_3]_2$ is made alkaline, $CaCO_3$ will be deposited. In warm tropical waters, as at the Bahama banks, $CaCO_3$ is being precipitated in great quantities.

Various investigators, particularly Molisch and Drew, attempted to implicate specific organisms in this process. However, any organism producing alkali would have the same effect. Precipitation of calcium carbonate from peptone media, as by Molisch's "*B. calciprecipitans*" is due merely to the ammonia resulting from deamination. Sulphate reduction also removes a strong anion and thus makes the solution alkaline; it too has been considered responsible for $CaCO_3$ precipitation (Bavendamm, 1932). There is evidently no reason to believe that any particular "chalk-precipitating" organisms are involved in the process.

Decomposition of Nitrous Oxide

So far as is known, N_2O does not occur in the atmosphere, although it is found in traces in soil; it must therefore be readily decomposed. *Ps. stutzeri* and *Ps. aeruginosa* rapidly reduce N_2O to N_2 (Beijerinck and Minkman, 1909); the former converted a gas containing 85 per cent N_2O to one containing

90 per cent N_2 in six days at 37° C. Denitrobacillus can convert 40 per cent of N_2O to N_2 in eight days at 30° C (Verhoeven, 1952). Oxidation of the reduced cytochrome *c* by N_2O is striking; the band at 556 m/i disappears (Sato and Egami, 1949). In adapted cells of *Ps. stutzeri*, and in Chung and Najjar's enzyme prepared from it, or in Verhoeven and Takeda's enzyme from *Ps. aeruginosa*, nitrate, nitrite, or N_2O causes it to disappear at once. The band will of course disappear in oxygen too.

Reduction of Nitrate to Ammonia

This is the most extreme reduction possible; however, it is not uncommon, having been recorded for Azotobacter, Radiobacter, *Cl. welchii* Desulfovibrio, Denitrobacillus, *Bac. subtilis* (the Marburg strain), and some strains of *E. coli*. *Cl. welchii* and *E. coli* strain Bn can use H_2 gas as reducer, thus:

$$HNO_3 + 4\ H_2 \rightarrow NH_3 + 3\ H_2O$$

The *E. coli* strain can also reduce nitrite, hyponitrite, and hydroxylamine to ammonia, using H_2. For this reason it can grow aerobically on any of these as sole nitrogen source, or anaerobically on lactate and nitrate (McNall and Atkinson, 1956, 1957). With Desulfovibrio, lactate was the hydrogen donor, and the rapid nitrate reduction did not interfere with the reduction of sulphate, which went on simultaneously (Baumann and Denk, 1951).

Nitrite is of course the first intermediate, and much evidence indicates that hydroxylamine is the last. With *Cl. welchii* it yields NH_3 more rapidly than nitrite or nitrate (Woods, 1938), and Desulfovibrio reduces it readily (Pichinoty and Senez, 1956). *Bac. pumilus* reduces NH_4OH about as well as nitrite, using reduced FAD or methylene blue as H donor (Taniguchi *et al.*, 1956). *E. coli* will reduce it, with glucose, in well-aerated cultures, and the strain *E. coli* B, which grows on nitrate or nitrite in an atmosphere of hydrogen, also can use NH_4OH or N_2O (McNall arid Atkinson, 1957). Small amounts of NH_4OH have been detected in denitrifying cultures (Verhoeven, 1952).

Strangely enough, aeration *promotes* the reduction to ammonia. Denitrobacillus, in shallow-layer cultures with glucose as H donor, gives a 40 to 90 per cent yield of ammonia (Verhoeven, 1952). *Bac. subtilis* does the same, and when fed $KN^{15}O_3$, the reduction is so rapid that within 45 minutes highly labelled $N^{16}H_S$ is found in the medium (Hall and MacVicar, 1955). Evidently, therefore, reduction to ammonia is a different process from the other denitrification reactions. It can be brought about by such organisms as *Bac. subtilis,* which do not produce N_2 or N_2O, and it may even be promoted by aeration. Kluyver (1953), who devoted several of his last years to the study of *Micrococcus denitrificans,* made the attractive suggestion that while N_2 requires the union of two molecules of nitrite (or the next intermediate), NH_3 does not; thus, the action of O_2 would be just to inhibit *dimerisation* rather than denitrification as a whole.

It must be remembered, too, that many organisms can readily grow on nitrate as sole nitrogen source. This means that nitrate must be reduced to amino acids, and hence probably to NH_3, in the small amounts needed for growth, by a wide variety of organisms. It seems that a sharp distinction must be made between *bulk* nitrate reduction, with accumulation of N_2O, NH_3 or other reduced products, and *assimilatory* nitrate reduction, in which the only product is cells. The first case, in which nitrate acts as terminal electron acceptor for all substrates, could be called nitrate respiration.

Agricultural Importance of Denitrification

It was the loss of nitrogen from soil that led Gayon and Dupetit (1882) to discover denitrification. Counts of denitrifying organisms show them to represent about 1 in 10,000 of the total soil bacteria; Diiggeli, for exampie, found 830 out of 8.4 million total count in garden soil, and 1720 out of 3.6 million in vineyard soil. Nitrifiers are probably present in similar numbers. Russell (1950) states that crops typically absorb only about one-half the nitrogen supplied them, while the rest is lost, partly by leaching and partly by denitrification.

Nitrate reduction is probably most active in the lower layers of ordinary soil or in waterlogged soils such as rice fields. In these, during the period when they are flooded, nitrate often disappears entirely and ammonia may accumulate. When the soil dries out again, nitrate reappears, in amount about equal to the ammonia, and the cycle continues. Waterlogged soils frequently contain appreciable amounts of nitrite, which may be toxic to the crop.

Liming of such soils, to bring the *pH.* up, may cause loss of nitrogen, probably through promoting activity of denitrifiers. Similarly, liming of (waterlogged) peat soils has been claimed to cause serious nitrogen loss by denitrification. Ricefield soils have been stated to lose nitrogen spontaneously.

Appreciable denitrification can occur aerobically also. Soil aerated in cylinders with added nitrite or nitrate shows a small utilisation of the added nitrogen for growth of microorganisms, but loses N when much organic matter is present. This means, in the first place, that a readily available hydrogen donor is needed for appreciable reduction to occur; the humus is too slowly decomposed to allow of much nitrate reduction. Second, it means that moderate aeration does not prevent denitrification any more in soil than in culture vessels. Important losses of nitrogen are claimed to take place.

Loss of Nitrogen during Aeration of Soils

These losses are no doubt partly due to leaching into the subsoil, but in part they are due to reduction to N_2, caused by the addition of fresh, decomposable organic matter. They have also been ascribed to photochemical decomposition of NH_4NO_2 (Dhar and Mukherji, 1935). The loss of nitrogen from standing manure heaps is mainly due, however, to volatilisation of ammonia, set free by urea organisms and by deamination; it is largely aerobic in nature, because in air the temperature of manure goes very high (up to 72° C), and this drives off "torrents of ammonia" (Gayon, 1884).

In experiments on sewage decomposing under aerobic

conditions (activated sludge process), a large loss, apparently as N_2, occurred when extra nitrite or nitrate was added (Corbet and Wooldridge, 1940). Activated sludge plants in which nitrification is occurring may, therefore, be subject to some (probably small) loss of nitrogen. It may well be that in addition to the losses of nitrogen in fires, burning of agricultural refuse, cremation, and the like, a very considerable drain on the nitrogen cycle is that due to nitrate reduction.

Role of Microorganisms

Nature of Proteins: Proteins consist of large numbers of α-amino acids linked together by the elimination of water. Since amino acids in neutral solution are not in the neutral form RCHCOOH but in the doubly charged form NH_2, $RCHCOO^-$, the resulting protein molecules are highly polar; they are therefore, soluble in water but not in organic solvents. They are usually rendered insoluble ("denatured") by heat. Some, like the fibroin of silk or the elastin of muscle ligaments, are long fibrous "macromolecules" of extremely low solubility; others, like ovalbumin of egg white, are more nearly spherical and highly water soluble. Their molecular weights range from 17,000 up to several millions. Many of them are enzymes, others reserve nutrients (as in eggs and seeds), others are structural (like collagen) or mechanical (like the proteins of flagella and muscles). The toxins of pathogenic bacteria (tetanus, diphtheria, botulism, etc.) are all proteins.

Proteins are normally composed of 21 different amino acids; however, one of these, hydroxyproline, is thought to be formed *in situ* from protein-bound proline, and cystine is readily formed from cysteine, so that only 19 molecular types are used in protein synthesis. All the amino acids participating in the synthesis have amino groups *a* to the carboxyl and are of the L-configuration.

Some D-amino acids occur in nature, especially in bacteria, e.g., in cell walls and capsules and in antibiotic substances, but they are not found in proteins. The L-form of all the constituent amino acids makes the proteins asymmetric in structure, the typical form being that of a helix (Pauling and Corey, 1951).

The composition of some selected proteins widely vary. Hydroxylysine is only found in a few; tryptophan and cysteine are completely missing from collagen (and therefore, from gelatin which is derived from it); histidine and methionine are absent from some strains of tobacco mosaic virus; glycine is relatively low in most proteins, but in collagen becomes by far the most important constituent; serine reaches a similar status in the fibroin of silk; and so on. In addition to the amino acids many proteins are linked to lipids, metal atoms, or biologically active "prosthetic groups," like hemin, riboflavin, or DPN.

With the exception of links between the S-atoms of cystine in the native keratins of skin and wool, and in insulin, all links between amino acids in the protein molecule are of the peptide type, resulting from the loss of the elements of water:

> This loss of water comes about initially through the formation of aminoacyl derivatives of adenylic acid; from these "activated amino acids" a series of reactions leads to protein synthesis. In the reverse process, the addition of the elements of water (hydrolysis) at the peptide bond causes protein breakdown.

In a few proteins, e.g., insulin and the enzyme ribonuclease, the exact sequence of all the amino acids has been determined; in many others, parts of the sequence are known. Two fundamental stages of proteolysis (= protein hydrolysis) may be distinguished:

1. The first is a breakdown into relatively large molecules, the polypeptides, each containing two or more amino acids. This is brought about by *proteinases*, i.e., enzymes that can hydrolyse a peptide bond when it is situated within a large molecule. (Most proteinases can also hydrolyse relatively small peptides or even ester linkages as well).

2. The breakdown of polypeptides, or of di- and tripeptides, to the free amino acids is brought about by various types of *peptidases.* These are enzymes that can attack a peptide bond only when one of the adjoining groups of the linked amino acids is free. Some require a free amino group (aminopeptidases), some a carboxyl group (carboxypeptidases).

The subsequent decomposition of the amino acids takes place in various ways, giving either ammonia, which is the main product, or in some instances other organic nitrogen compounds.

Attack Upon Proteins

Native Protein: Purified proteins, alone or with the usual inorganic salts, seldom provide a good medium for the growth of bacteria, even of organisms that are in general actively proteolytic. Crystalline ovalbumin, serum albumin, and serum globulin, together with inorganic salts, allow practically no growth of organisms of the coliform group or of *M. aureus* or Gonococcus, and very little even of highly proteolylic organisms such as *Bac. subtilis, B. prodigiosum,* and *Cl. pulrificum* (Sperry and Rettger, 1915).

In both cases, however, the addition of peptone (the mixture of polypeptides obtained by treating proteins with pepsin) allowed vigorous growth to take place. With Proteus and others (but not with *E. coli* or *S. typhosd),* this growth was accompanied by active proteolysis. Thus, it seems that the proteinases of many bacteria are elaborated only when vigorous growth takes place, and that some nitrogen source other than pure protein is needed to start the growth. This is particularly true with native, or undenatured, proteins.

Streptomyces griseus, which produces streptomycin, excretes a powerful proteinase which does attack egg albumin. The crystalline enzyme also attacks polypeptides and even dipeptides (Nomoto *et al.,* 1959). At the opposite extreme is *Bac. subtil is,* which attacks purified ovalbumin without causing extensive hydrolysis. This organism, or the crystalline enzyme "subtilisin" prepared from it (Guntelberg and Ottesen, 1952; Ottesen, 1957), merely splits off six to seven amino acids to produce a new protein, called plakalbumin (because it crystallises in plates). Except for minor differences in solubility and in immunological properties (Grabar and Kaminski, 1950), the plakalbumin still very strongly resembles the original ovalbumin. One of the few native proteins readily attacked by bacteria is collagen, the protein of bones, skin, and animal connective tissue.

Three aerobes, *Bac. brevis, mycoides,* and *mesentericus,* and several anaerobes, including *Cl. histolyticum, welchii,* and *lentoputrescens,* hydrolyse native beef tendon or purified collagen (Jennison, 1945). Some 30 strains of *Cl. welchii* hydrolyse hide powder rapidly, and *Cl. histolyticum* is even more active (Evans, 1948). The effective enzyme, *collagenase,* is excreted by these Clostridia and is responsible for their ability to break down living tissue; the enzyme from *Cl. welchii* solubilises purified collagen in a few hours at 38° C (Delaunay *et al.*, 1949). It is this action that sometimes turns Clostridium-infected wounds into a semi-liquid state.

The enzyme comprises part of the mixture of toxic substances formed by the most potent type of *Cl. welchii,* type A (Bidwell and van Heyningen, 1948; Bid well, 1949- 1950). Types B and D of this organism, which cannot break down living tissues, do not contain collagenase, but type B instead has an enzyme that hydrolyses only denatured animal proteins and gelatin (Bidwell, 1950). Several studies of the purified enzyme have been made, especially from *Cl. histolyticum.* Grant and Alburn (1959) have identified at least three separate collagenases from a pure culture filtrate, two with optimum *pH.* about 8, and one near 7. All are activated by Ca and attack gelatin rapidly, but other proteins only slowly. A non-specific enzyme attacking several denatured proteins was present as well.

Elastin, the fibrous protein of ligaments, is not attacked by collagenase, and only a few bacteria can hydrolyse it. Besides *Vibrio comma* and a Pseudomonas (Narayanan *et al.*, 1953), one of the few other bacteria attacking it is a Flavobacterium isolated from disease pockets around teeth (Mandl and Cohen, 1960). Its enzyme (elastase), after purification, attacked no other substrate. It had a *pH* optimum of 7.5 and was inhibited by Zn and Cd but not by other metals. Elastase has also been reported in five Actinomycetes (Sbarra *et al.*, 1960).

Keratin, the protein of hair, wool, horn, and feathers, is also highly resistant to most bacteria, although attacked by certain insects, including of course the clothes moth. Much of its stability is due to S-S linkages between amino acid chains; when these are reduced, the rest of the molecule is more hydrolysable, and thus keratin hydrolysis depends in the first stage on reduction.

For this reason, keratin that has been autoclaved or ground is sometimes attacked by organisms that cannot attack the native keratin, since these treatments may reduce the S-S linkages. The more careful studies have shown that truly native keratin (such as defatted wool or hair) is attacked only by: (1) skin-disease fungi or *dermatophytes*, especially *Microsporum gypseum* whose ability to attack native keratin has been considered a diagnostic test (Vanbreuseghem, 1952); (2) a few Streptomycetes, among which *Str. fradiae* is very active, solubilising 90 per cent of the keratin from various sources in about a week at 28° C. The S-S bonds were broken down to SH compounds, and much of the nitrogen liberated as ammonia (Noval and Nickerson, 1959). Some keratin-digesting activity, promoted by metallic ions, was excreted into the medium. Six other Streptomycetes and two Nocardias had some action on wool. The enzyme of *Str. fradiae* occurs conjugated to an acidic polymer, from which, however, it can be separated and crystallised (Nickerson *et al.*, 1962). The enzyme has maximal activity at its isoelectric point, which lies at *pH*. 9.

That members of the Actinomycetacae should be the major keratin decomposers in nature agrees with observations on hair and wool fragments in soil, which by direct microscopy show the presence of many attached Actinomycetes, possibly *Actinoplanes* species (Gaertner, 1955).

Bacterial Protein

An interesting substrate is the protein of the bacterial cell itself. Why proteinases do not attack the proteins of the cells that produce them is still a mystery, but it is not more mysterious with bacteria than with any other cells. Perhaps the formation of a proteinase automatically requires the formation of a specific antagonist for it, which remains within the cell when the proteinase diffuses out. However, this may be, it is enough to kill bacteria by heat in order to make them sensitive to proteolytic enzymes.

Thus, heat-killed pneumococci are digested by trypsin (Dubos, 1937); heat-killed Aerobacter by an enzyme from a soil Actinomycete (Muggleton and Webb, 1952); and heat-killed *M. aureus* and *E. coli* by an enzyme from a *Streptomyces* species (Tai and van Heyningen, 1951; Born, 1952). Living cells are not so attacked. Tai and van Heyningen (1951) have shown that the Streptomyces excretes two enzymes, one active only on casein and the other on the *E. coli* protein, and that the latter is strongly absorbed by the cells that it can lyse. It appears that simple heat-killing does not suffice to make the cells lysable nor to make them adsorb the enzyme, since heating 10 minutes at 55° C (which kills virtually all), starving them to death for 12 days in distilled water, or killing by antibiotics is ineffective; heating 15 minutes at 75° C or higher is required. Evidently the protein must be extensively denatured.

These lytic enzymes in some ways resemble trypsin; they are activated by Mn and other metals and inhibited by H_2S and other compounds and have optima about pH, 7. Trypsin, indeed, has similar effects and also acts only on heat-killed cells. The enzymes do not attack the cell walls, and when their lytic action is complete, the residual walls can be seen more or less intact (Salton, 1953, 1955). An enzyme in *Bac. cereus* spores that lyses the *walls* of the vegetative cells (Strange and Dark, 1957) is probably not a proteinase. Gram-positive cells are lysed very much more slowly than gram- negative ones, both by trypsin and by the Streptomyces enzymes. *E. coli, Ps. fluorescens, Sp. serpens,* and *Aerob. aerogenes* are sharply distinguished in this respect from *M. aureus, Bac. megaterium;* and *Cl. welchii;* the gram-staining complex apparently protects the cell protein against the enzyme, because *Cl. welchii* cells that have been made gram-negative by autolysis or by extraction with cholate become more sensitive to proteolysis (Muggleton and Webb, 1952). By contrast, the cytoplasm of yeasts, fungi, and even some bacilli is lysed readily by papain (Gorbach and Hoffmann, 1956).

Denatured Protein

Since casein and gelatin are not native proteins, they are easily attacked. For instance, the proteinase of yeast attacks casein, gelatin, and denatured ovalbumin, but not native ovalbumin. The hydrolysis of casein by *Penicillium camemberti* is an important part of the ripening and softening of Camembert cheese. It is common bacteriological practice to determine the ability of organisms to hydrolyse gelatin and to take this as a measure of proteolytic activity.

The hydrolysis can be detected by observing the clearing of agar containing casein or gelatin (Dion, 1950) but more commonly by inoculating tubes of gelatin medium. A medium containing peptone, salts, and about 10 per cent gelatin is filled deep into test tubes and inoculated by stabbing with a straight needle previously touched with the culture. Evidently these indicate growth and proteolysis, of types: (1) facultatively aerobic or anaerobic, (2) strictly aerobic, and (3) principally anaerobic. If the time, temperature, and composition of the medium are standard, and the stab is carefully made so that the gel closes up again on removing the needle, the results have considerable relative value.

The formation of gelatinase by several bacteria is strongly

promoted by calcium (Haines, 1932, 1933); 0.01 M $CaCl_2$, or less, is sufficient (Gorini and Fromageot, 1949). With fungi or Streptomycetes, Mg is often more effective than Ca (Dion, 1950). Sugars, on the other hand, inhibit formation of the enzyme, 1.5 per cent glucose being enough to render *Proteus vulgaris* completely inactive (Kendall, 1923). Similarly, protein hydrolysis by the bacteria of cattle rumen is strongly inhibited by starch, which is itself rapidly metabolised (Warner, 1956). This phenomenon seems not to occur in fungi, whose proteinase formation is sometimes increased by sugars. Thus, *Gliocladium roseum* gives increasing yields of proteinase with glucose, up to a maximum at 8 per cent glucose (Dion, 1950); perhaps this is due to the promotion of its growth.

Attack Upon Peptides

The ability to grow on peptone, proteose, and other partially hydro- lysed proteins is common to most organisms. Even bacteria like *E. coli,* which do not attack proteins, will grow well on peptones. Streptococci grow better when given arginine peptides than on arginine itself, because they decompose the amino acid (to ornithine) too rapidly (Gale, 1945). The same is true for peptides of tyrosine.

The peptidases of bacteria have not been extensively studied. Those of the lactic bacteria have *pH* optima at 5.5 to 6 or 6.8, while those from four gram-negative rods, three bacilli, and three clostridia have optima between *pH 8* and 9 (Berger *et al.*, 1938, 1941). The polypeptidase of yeast and the crystalline proteinase-peptidase of *Str. griseus* (Nomoto *et al.*, 1959) are both optimal between *pH, 7* and 8. Several bacterial peptidases are activated by Fe or Mn (Maschmann, 1944), thus resembling the Mn- or Co-requiring enzymes of animal tissues, while four gram-negative rods *(E. coli, Proteus, Ps. fluorescens,* and *Ps.* yielded peptidases activated by Mg (Berger *et al.*, 1938), as is the erepsin of animal intestines.

The Mycobacteria yield peptidases readily, *Mycob. smegmatis* excreting it into the medium, while *Mycob. tuberculosis* yields them on grinding the cells (Roulet and Zeller, 1948). Glycyl-L-leucine was the most sensitive substrate.

The lactic bacterium *Leuconostoc mesenteroides* has at least two di- and two tripeptidases, activated by metal ions, which can hydrolyse peptides of D-amino acids as fast as those of L-amino acids, in contract to most animal peptidases which are restricted to the L-isomers (Berger, *et al.*, 1938). It is evident that bacterial peptide hydrolysis requires much more patient study.

Nature of Enzymes

In animal tissues, where proteolytic enzymes have been the most thoroughly studied, four types are recognised:

1. *Pepsin,* active only at a very acid *pH.* and destroyed by alkalies, attacking some native proteins including haemoglobin, and clotting milk. Pepsin splits peptides if they contain an aromatic amino acid. Pepsin is liberated by acid from its inactive precursor in the stomach, pepsinogen.
2. *Trypsin,* optimum *pH,* 8, attacking native proteins only weakly, but denatured or modified proteins vigorously; it also clots milk. Trypsin can attack amide or ester linkages of peptides with a free basic group such as benzoylarginine amide or benzoylar- ginine methyl ester (Schwert *et al.,* 1948, and earlier work there cited). In the need for a free basic group and a covered carboxyl, the *aminopeptidases* resemble trypsin. The activity of trypsin is increased by divalent metal ions and inhibited by cysteine. It occurs as the in- active form, trypsinogen, which is made active by a special enzyme, enterokinase; it also converts itself autocatalytically at *pH,* 8.
3. *Chymotrypsin,* most active at *pH, 7* to 8, attacking few native proteins but many peptides, natural and synthetic. Chymotrypsin has no requirement for basic or acidic groups, but does require an aromatic ring

(or a methionine residue). It occurs in the intestine as an inactive precursor, chymotrypsinogen, which is converted to the active form by trypsin.

4. *Papain* and some cathepsins, ficin, bromelin, etc., a group of enzymes found in liver, spleen, and in higher plants, optimum *pH* about 5, requiring to be activated by HCN, H_2S groups. Papain will attack tripeptides. Like trypsin, it can clot milk. The bacterial proteinases have some similarities to these animal enzymes. In one case, the similarity extends to the formation of enzyme from inactive precursor, since a micrococcus has been found to excrete a compound that is autocatalytically converted to a proteinase on standing a few hours (Gorini and Lanzavecchia, 1954). In general, proteinases must of course be excreted from the cell, since proteins cannot pass the permeability barrier to be attacked inside. It is true that the cell's own proteins must also be broken down in normal metabolism, and indeed in non-growing cells, this happens actively, leading to rapid "turnover," as has been shown with *E. coli* by following the fate of selected C^{14}-labelled amino acids (Mandelstam and Halvorson, 1960).

During growth, this proteolysis is in some way prevented. While these *endoproteinases* are sometimes excreted, especially in old cultures (Berger, *et al.* 1938), the typical proteolytic bacteria, like *B. prodigiosum* or *Ps. fluorescens,* excrete proteinases vigorously from the start of growth (Virtanen and Tarnaneii, 1932; Maschmann, 1937- 1938).

The peptidases are more commonly endoenzymes, which can only be extracted by grinding, autolysing, or repeated freezing and thawing (Berger *et al.*, 1938; Roulet and Zeller, 1948), although they are excreted by *Bac. megaterium.* The two enzyme types thus show a natural separation; for example, *Ps. fluorescens* actively excretes a proteinase during growth, while from the cells, a polypeptidase and a dipeptidase could be extracted (Virtanen and Tarnanen, 1932). Several Clostridia

(CL welchii, Cl. chauvoei, Cl. histolyticum) excrete an enzyme of the papain type, which is activated by cysteine, and hydrolyses the basic protein clupeine (Maschmann, 1937-1938). The proteinases of other clostridia and of Proteus are also activated by cysteine (Weil *et al.*, 1939).

The proteinase that has been crystallised from group A streptococci requires activation by HCN, cysteine, etc., and in this resembles papain (Mycek *et al.*$_9$ 1952), but its action on synthetic substrates such as the amides of benzoyl-L-arginine and ben- zoyl-L-histidine rather resembles that of trypsin. The collagenase of *Cl. histolyticum,* which is the most active of all collagenases and has been extensively purified though not crystallised, is activated both by cysteine and by $FeSO_4$. Several aerobes have proteinases that are inhibited by cysteine but activated by Fe^{++} or by cysteine plus Fe^{++} (Maschmann, 1944).

The enzymes hydrolysing bacterial proteins, resemble trypsin in many ways. So does an enzyme from *Bac. subtilis,* which is sold for industrial use in Japan under the name of "Nagarse." This has been crystallised by Hagihara *et al.* (1958) and found more heat-stable and more resistant to natural inhibitors than is crystalline trypsin.

However, it has long been known that all hydrolytic enzymes are sensitive to this reagent, because the amino acid serine is an essential part of the active hydrolytic centre, and diisofluorophosphate reacts with the OH of serine.

This property is therefore, not critical. Recently three crystalline enzymes from bacilli, namely, a proteinase from *Bac. cereus,* the Nagarse enzyme, and Ottesen's subtilisin, have been carefully compared (Wieland *et al.*, 1960). All three have *pH* optima 7 to 7.5, coagulate milk (rennin action), and inactivate lactic dehydrogenase at the same rate. They have the same sedimentation constant in the ultracentrifuge, the same amino acid composition, and even the same amino acid at the end of the chain, namely, alanine. Apparently, therefore, they are identical or almost so.

In the pathogenic clostridia, it is the variety rather than the similarities that is notable, because there are enzymes of at least four types (Maschmann, 1937-1938); (1) a collagenase that also acts on gelatin, (2) a gelatinase, (3) a less specific proteinase that acts on gelatin, casein, fibrin, and ovalbumin *(CL botulinum)*, and (4) the papain-like enzyme that attacks clupeine.

A number of peptidases are present as well. The *pH*, optima offer limited clues to the nature of the enzymes. The gelatinases from six organisms all have their optima close to *pH, 7* (Maschmann, 1937-1938; Haines, 1933); the proteinase of *Lactob. casei* has optimum close to *pH*, 6, as do the gelatinases of several clostridia.

The collagenase of *Cl. welchii* has a very narrow *pH* band of action with optimum near 7 (Bidwell, 1950). A number of imperfect fungi form proteinases which have pH optima between 7 and 8, are unstable below *pH*. 5, and generally are more or less of the trypsin type (Me- Connell, 1950). The enzymes from Trichophyton and other dermato- phytes also have optima near 8 (Tate, 1929). On the other hand, yeast proteinase has its optimum at *pH*, 5. The proteinase of *Aspergillus oryzae*, which has been crystallised (Crewther and Lennox, 1950), has potential industrial uses, like the Japanese preparation from *Bac. cereus* of Hagihara *et al.* (1958); industry has shown much interest in these enzymes.

Ammonification in Soil

The final products of proteolysis are the amino acids. Bacterial attack on these substances involves reactions of several types, but in the main the nitrogenous product is ammonia. The study of amino acid breakdown is, therefore, in good part the study of ammonia production by bacteria or by their enzymes. It was a very early observation that nitrogenous material added to soil readily gives rise to ammonia. Marchal, however (1894), showed that sterilised soil gave no ammonia when mixed with nitrogenous matter, so that the process is evidently bacterial. The production of NH_3 from faeces is also bacterial (Severin, 1895-1901), about 40 per cent of human faeces consisting of bacteria. Of a number of organisms grown upon unpurified ovalbumin, Marchal found that *Bac. mycoides* was the most active ammonifier, converting 40 per cent of the added protein nitrogen to ammonia in 15 days at 30° C. However, *Proteus vulgaris,* an Actinomyces species, and many other bacteria, Aspergillus, Cephalothecium, and other fungi, also produced ammonia; he recognised, therefore, that the faculty of decomposing nitrogenous compounds to ammonia is widely distributed.

In the soil, ammonia production is often the result of collusion between different organisms, some being mainly responsible for the initial hydrolysis of polymers such as proteins and aminopolysaccharides, while others rapidly liberate ammonia from the resulting smaller molecules.

Of the main paths of amino acid breakdown, that depending on oxidation is the most important and will be considered first.

Oxidative Deamination

Process itself: Marchal appreciated that ammonification in soil is essentially oxidation, carbon dioxide and water being formed at the same time and in approximately the amounts corresponding to complete oxidation. The first step in this process, in which ammonia is eliminated, is called oxidative deamination. According to the researches of Knoop and of Neubauer on animal tissue, it proceeds in two stages dehydrogenation to the imino acid, and subsequent hydrolysis to the keto acid:

$$\text{COOH COOH COOH}$$

$$NH_2\ NH + H_2O\ O + NH_3$$

Oxidative deamination is probably the commonest type of amino acid breakdown. Like so many biological oxidations, it may be imitated by quinones (Schaaf, 1929; Kisch, 1931-1932) and other compounds containing two carbonyl groups, as well as by H_2O_2.

Unfortunately, oxidative deamination by microorganisms is nearly always accompanied by secondary reactions, so that the highly reactive keto acids seldom appear. The evidence for its widespread occurrence is therefore, in good part indirect, but four types of evidence combine to make it convincing. Most of this was obtained long ago.

1. Deamination by washed bacteria is often strictly oxidative; Janke and Tayenthal (1937), using *E. coli* and *Bac. mycoides,* found glycine to be attacked only in air, and Stephenson and Gale (1937) found the same for alanine and glutamic acids. In favourable cases, as when *Proteus vulgaris* acts on histidine (Raistrick, 1919), leu cine, phenylalanine, or methionine (Bernheim *et al.*, 1935), only one atom of O_2 is absorbed and one mole NH_3 produced.
2. Where the products are not the corresponding keto acids, they are substances that could have been formed readily from these keto acids. For example, glutamic acid is readily converted to succinic acid by yeast, but α-ketoglutaric acid was converted to succinic acid in a 99 per cent yield (Neuberg and Ringer, 1915, 1918). Similarly, alanine is oxidised to acetic acid by many organisms, but the intermediate keto acid, pyruvic acid, is oxidised to acetic acid even more rapidly. *Proteus vulgaris* also rapidly oxidises α-keto butyric acid (Stumpf and Green, 1944).
3. In favourable circumstances, the keto acid has actually been isolated. Neubauer and Fromherz (1911), realising that with natural amino acids the resulting keto acid is readily broken down, chose an amino acid that does not occur naturally, namely, phenylglycine. By exposing this to yeast (in the presence of sugar), they obtained phenyl- glyoxylic acid, isolated in 20 per cent yield:

$$C_6H_5CHNH_2COOH \longrightarrow C_6H_6COCOOH + NH_3$$

Similarly, glycine exposed to washed cells of *E. coli* or

Bac. mycoides gave rise to glyoxylic acid (Janke and Tayenthal, 1937):

$$CH_2NH_2COOH \longrightarrow CHOCOOH + NH_3$$

but some CO_2 was evolved as well, indicating more extensive breakdown. With fungi *(Asp. oryzae)* small yields of phenylpyruvic acid were obtained from phenylalanine, and of isopropylpyruvic acid from leucine (Uyemura, 1937); *Fusarium lini* also converts alanine to pyruvic acid (Wirth and Nord, 1943).

4. Cell-free preparations, from which some at least of the accompanying enzymes have been removed, give cleaner results. In one such preparation from *Proteus vulgaris*, the keto acids were only very slowly oxidised, if at all, and the theoretical requirements, i.e., equimolecular quantities of oxygen consumed and of NH_3 and keto acid produced, were realised. With this same preparation, the formation of keto acids from 11 different amino acids was demonstrated. In the special case of glutamic acid, the dehydrogenation is carried out with a hydrogen acceptor other than oxygen; in animal tissues, in Treponema, and in *Cl. sporogenes* this is DPN (= NAD) (Olson and Anfinsen, 1952-1953); in *E. coli* and probably most bacteria, it is TPN (= NADP) (Adler *et al.*, 1938), while in yeast both a DPN- linked and a TPN-linked enzyme are present (Holzer and Schneider, 1957). The first step, with the ensuing hydrolysis, is given by equations (1) and (2). The reduced co-enzyme reduces flavoprotein, and the reduced flavoprotein is reoxidised by oxygen.

Formal proof that it is ammonium ion, NH_4^+ that participates, rather than NIX,, has been given by Fisher and McGregor (1960) from a study of pH sensitivity. Accordingly, all the participants have been shown in their corresponding charged forms. Neurospora has several mutants that are deficient in nitrogen assimilation. Study of these has brought to light three different glutamic dehydrogenases (Fincham, 1960). One is like that of bacteria, one is activated by a few minutes at 35° C and loses its activity slowly on cooling, and the third differs in having a much higher affinity for all its substrates, especially for ammonia. They are inherited independently.

All three use TPN (= NADP) as hydrogen acceptor and are otherwise superficially alike. Thus, what appears to be a single enzyme can occur in various modifications, and one can only suppose that when other organisms receive equally detailed study, many similar cases will come to light.

A second special case is that of L-alanine. This deaminase, or dehydrogenase, has so far only been found in *Bac. subtilis* and other bacilli; the equilibrium is in the direction of alanine synthesis, and apparently for these organisms the reaction is the main path of ammonia assimilation (Hong *et al.*, 1959). In other bacteria this function is served by glutamic dehydrogenase, which thus has a much greater importance than merely *as a* deaminase for one amino acid. The L- alanine dehydrogenase of *Bac. cereus* is specific for DPN and has *pH* optimum 8.8 (for animation) and 9.8 (for deamination), values that allow it to function well in presence of free ammonia (O'Connor and Halvorson, 1960).

In view of the close relation between the functions of these two dehydrogenases in amino acid metabolism, it is truly remarkable that the enzymes themselves are just as closely related. For recent work shows that the glutamic dehydrogenase molecule readily dissociates into four subunits (Frieden, 1959), and one of these acts as alanine dehydrogenase (Tomkins *et al.*, 1961). The dissociation is brought about by certain steroids and metal-chelators, while reassociation is favoured by ADP, DPN, and TPN; correspondingly, the first group of reagents inhibits activity on glutamate and promotes activity on alanine, while the second group has the opposite effect.

There is some evidence for a special deaminase for phenylalanine, too, since Proteus characteristically oxidises it faster than other bacteria. In this case, the keto acid, phenylpyruvic acid, is readily identified as the product because of its colour reaction with $FeCl_3$ (Henriksen, 1950). The ability to produce this reaction in phenylalanine broth has, therefore, been proposed as a diagnostic test for Proteus; it can be done in 10 minutes (Ben Hamida and le Minor, 1956). Other amino acids are deaminated directly by a pair of flavoprotein enzymes, the *amino acid oxidases,* whose only specificity is for the L-isomers and the D-isomers, respectively. Some organisms possess only, or mainly, one type, e.g., *Ps. aeruginosa* does not attack D-leucine, D-isoleucine, or D- histidine, and although it oxidises some other o-forms, they are not deaminated (Webster and Bernheim, 1936). In *E. coli,* o-glutamate is attacked only one-fifth as rapidly as the L-form, while in *Proteus vulgaris,* the enzyme for the L-isomers is the only one present in the cell-free preparation. However, in animal tissue, the D-forms are generally the more rapidly deaminated, and in several fungi very active D-amino acid oxidases are present (Emerson *et al.,* 1950).

Optimum *pH.* for oxidative deamination, either in vivo or in vitro, is between 6.5 and 8. When the flavoprotein is reoxidised by O_2, the overall free energy yield is about 20,000 calories per mole, which means that the reaction goes fairly completely. The status of the imino acid as the primary product, although generally accepted for some time, has recently been given direct chemical support, leucine and the L-amino acid oxidase being used (Frieden and Velick, 1957). A peculiar feature is that glucose in the culture medium inhibits formation of these enzymes, although glucose does not affect the deamination process itself (Stephenson and Gale, 1937).

5. Where oxidative deamination is reversed, i.e., in *reductive animation*, the role of the keto acid is clear. In preparations from yeast or *E. coli*, the reduction of TPN by glutamic acid is at once reversed by adding α-ketoglutaric acid, especially if NH_4^+ is added too.

   Similarly, the animation of pyruvic acid to alanine, first observed in fermenting yeast (Fromageot and Desnuelle, 1933-1934), can be readily carried out with L-alanine dehydrogenase; the enzyme of *Bac. cereus* and its spores aminates pyruvate about 10 times as fast as it deaminates alanine (O'Connor and Halvorson, 1960). Here the equilibrium is even farther towards synthesis. This fact is, as mentioned above, important because these two dehydrogenases appear to provide *the main pathways of ammonia assimilation*, the glutamic system for yeast and most bacteria, the alanine system for a few mutants of these and for the bacilli (Hong *et al.*, 1959). The general importance of glutamic acid as a route of nitrogen assimilation can be shown dramatically by following the appearance of C^{14}-compounds in yeast treated with C^{14}- ethanol (Aubert and Milhaud, 1956). Labelled glutamic acid appears in the cells within five seconds, aspartic acid after one minute, alanine and other amino acids only after 15 minutes. When the yeast has been grown on alcohol, the inter con versions are faster, but the order is the same and it is clear that glutamic acid is formed with great speed. Although the keto acid is optically inactive, its animation generally leads to the L-amino acid. Thus, the keto-analogues of leucine and valine, i.e., α-keto-isocaproic and a-keto-isovaleric acids, are reductively aminated by *Bac. subtilis* to L-leucine and L-valine, with less than 1 per cent of the D-isomer (Shah and King, 1958). (The formation of DL-alanine from pyruvate by this bacillus is an exception due to presence of a *racemose,* which brings D- and L-alanine into equilibrium.) The asymmetric formation of the amino acid must mean that the keto acid is rigidly held on the enzyme at three points while the ammonia reacts with it.

6. The relationship to keto acids is shown also by growth requirements. A Neurospora mutant requiring leucine for growth would grow immediately on α -keto-isocaproic arid, and another requiring valine would grow on α -keto-isovaleric acid (Bonner *el al.*, 1943). Aerobacter and *E. coli* mutants requiring valine both similarly respond to the keto acid, while another *E. coli* mutant actually accumulates the keto acid and is therefore, deficient in the animating system (Umbarger and Magasanik, 1951; Maas and Vogel, 1953).

Transamination

A special kind of reversible deamination leads to the formation, not of free NH_3, but of another amino acid. Alanine, for instance, can be deaminated to pyruvic acid in the presence of α -ketoglutaric acid.

This process, *transamination,* is of general importance. At first (Braun- stein and Kitzmann, 1937) it was thought to take place with only very few amino acids, but it is now known that virtually all amino acids take part. For example, dried cells of *E. coli* and *Ps. fluorescens* can each use α -ketoglutarate as an amino acceptor for 13 amino acids (Feld- man and Gunsalus, 1950); an extract from *Lactob. arabinosus* uses it for 16 (Meister, 1952). In *E. coli* also, α -keto-isovaleric acid acts as an amino acceptor for alanine and a-amino butyric acid, and in Brucella pyruvate is an amino acceptor for leucine. The role of aspartic acid as an amino group donor has been given special prominence in Virtanen's interpretation of the nitrogen fixation process in nodules and has been recorded in large numbers of microorganisms.

Transamination of the y-amino group of ornithine is an interesting case, because it yields glutamic y-semi-aldehyde (Fincham, 1953).

Since this aldehyde is a direct precursor of proline (Vogel and Davis, 1952), this reaction interconnects arginine breakdown with proline synthesis. Several other a-keto acids act as amino acceptors with ornithine. A parallel case is the transamination of glyoxylic acid from glutamic acid to yield glycine; the equilibrium here is in the direction of glycine formation (Metzler *et al.*, 1954).

Even D-amino acids can transaminate; *Bac. subtilis* uses a-ketoglutarate as an amino acceptor for D-alanine, forming D-glutamate. In view of the prevalence of D-glutamate polymers in capsules and elsewhere, this reaction is doubtless of great importance in bacteria. D-Alanine is produced by the racemase mentioned above.

With all these reactions, pyridoxal phosphate is needed as a co-enzyme. The apoenzyme (i.e., the protein part) is not a separate one for each amino acid, but like the amino acid oxidises above, is rather unspecific. In *E. coli*, there is evidence for only about three separate enzymes, each of which catalyses a large group of transaminations.

The peculiar readiness with which α -ketoglutarate acts as an amino acceptor, so that glutamic acid will be continually formed in a metabolising cell, lends special significance to the wide occurrence of glutamic dehydrogenase in bacteria. The two reactions may constitute a major route for the deamination of other amino acids, namely:

Amino acid + α -Ketoglutarate $\longrightarrow$ Keto acid + Glutamate

Glutamate + H_2O + TPN^+ $\rightarrow$ TPNH + H^+ + α -Ketoglutarate + NH_2,

The glutamic acid thus acts as a catalyst. Evidently this pathway is an alternative one to that using the amino acid oxidases. From what was said above, the pathway is a main one in the reverse direction too, glutamate acting as catalyst in the bacterial synthesis of many amino acids.

Secondary Reactions of the Keto Acids

1. *Reduction.* The keto acid may be directly reduced to the hydroxy acid:

$$2(H)\ R—CO—COOH \longrightarrow R—CHOH—COOH$$

For example, tyrosine in presence of glycerol and phosphate was converted by *E. coli* and *Proteus vulgaris* to p-hydroxy-phenyl-lactic acid (Sasaki, 1914). In this case R = $HOC_{,5}H_4$. Similarly, phenylglycine added to a yeast fermentation, in the presence of sugar, was converted to the hydroxy acid (mandelic) as well as partly to the keto acid.

Many other examples could be adduced. There is no evidence at all that organisms produce hydroxy acids from amino acids by direct *hydrolysis*, i.e.:

$$R\text{-}CHNH_2\text{-}COOH + H_2O \longrightarrow R\text{-}CHOH\text{-}COOH + NH_3$$

Where hydroxy acids appear, there is every reason to believe they are formed by oxidative deamination and subsequent reduction of the keto acid. Hydrolytic deamination appears to be restricted to amides.

2. *Decarboxylation.* In addition to oxidative decarboxylation keto acids can be "simply" decarboxylated to produce the aldehyde with one less carbon atom:

$$R\text{-}CO\text{-}COOH \longrightarrow CO_2 + R\text{-}CHO$$

Among those rapidly decarboxylated by the enzyme from yeast are the α-keto derivatives of isocaproic and isovaleric acids and of ethyl glutarate (Meister, 1952). All of these could arise by deamination of amino acids.

The resulting aldehyde can react in two ways:

1. It may be dehydrogenated to an acid, RCOOH, having one carbon atom less than the original amino acid. The overall reaction is thus not different from oxidative decarboxylation. In the few cases where aldehydes are identifiable intermediately, they are subject either to direct dehydrogenation or to dismutatiori; thus, acetaldehyde dehydrogenase forms acetic acid; its "mutase" forms a mixture of acetic acid and ethanol.
2. The alternative fate is that the aldehyde may become reduced to the alcohol:

   $$RCHO + 2(H) \longrightarrow RCH_2OH$$

   This reaction was first studied by Ehrlich in attempting to discover the source of the "fossil oils" (amyl alcohol and its isomers) produced in yeast fermentation. Esters of amyl alcohol are responsible for a good deal of the aroma or "bouquet" of fine wines. According to Heinzel- mann and Dehnicke (1915), who made an extended study of the conditions, fuel oils are produced by all yeasts; the amount is greater with grain mash than with potato mash, doubtless due to the high-protein content of the grains. Ehrlich (1907, 1909) showed that isoamyl alcohol arises from leucine:

   $$(CH_3)_2CHCH_2CHNH_2COOH + H_2O \longrightarrow (CH_3)_2CHCH_2CH_2OH + NH_3 + CO_2$$

Leucine Isoamyl Alcohol

Several fungi that carry out an alcoholic fermentation give the same reaction (Pringsheim, 1908); added leucine is quantitatively converted to isoamyl alcohol. Valine is converted to isobutyl alcohol (Ehrlich, 1906):

$$(CH_3)_2CH\text{-}CHNH_2COOH + H_2O \longrightarrow (CH_3)_2CH\text{-}CH_2OH + NH_3 + CO_2$$

and serine to ethylene glycol:

$$CH_2OHCHNH_2COOH + H_2O \longrightarrow CH_2OH\text{-}CH_2OH + NH_3 + CO_2$$

Phenylalanine and tryptophan (indolealanine) give, analogously, phenyl-ethanol and tryptophol (indole-ethanol) (Ehrlich, 1907, 1912), while phenylglycine gives benzyl alcohol (Neubauer and Fromherz, 1911). A cell-free preparation from yeast converts tyrosine to p-hydroxyphenylethanol (tyrosol), and here the presence of the intermediate p-hydroxy-phenylpyruvic acid was shown (Shanmuganathan and Elsden, 1958). Asparagine or ammonium sulphate prevent fuel oil production, probably by reaminating the intermediates. However, fossil oil is not produced only from external amino acids, since it can also be endogenous. Yeast suspended in glucose solution produces fossil oil at a slow but steady rate (Genevois and Lafon, 1957), and some of its constituents, e.g., ra-propanol and butanol, are not directly related to amino acids. Apparently the fusel oils can be produced from glucose by the same pathways as are used for the amino acids, since a yeast mutant that requires leucine for growth produced a fusel oil containing almost no isoamyl alcohol, and one requiring valine yielded very little isobutyl alcohol (Ingraham and Guymon, 1960). No doubt the synthesis goes as far as the related keto acid, and since the animation of this is genetically blocked in the mutant so that it cannot be converted to an amino acid, it is broken down directly; when a particular pathway is missing, both the amino acid and the fusel oil are affected.

3. *Oxidation:* Ketoglutarate is readily oxidised to succinale, and pyruvate to acetate, so that amino acids frequently give rise to an acid with one less carbon atom. Alanine yields acetic acid, and in a parallel reaction indolealanine (tryptophan) yields indoleacetic acid. This reaction, discovered (with mixed cultures) in 1880 by Salkowski, is now important on account of the role of indoleacetic acid as a plant growth hormone. With

Rhizopus suinus (a Phycomycete) the yield of indoleacetic acid was found approximately proportional to the extent of aeration of the culture (Thimann, 1935).

In several bacteria, the reaction requires a keto acid, an indication that the first step is a transamination rather than a direct oxidation; pyridoxal phosphate is needed too (Stowe, 1959). With *Acetobacter xylinum* both indoleacetic acid and indole-ethanol are formed, which implicates indole-acetaldehyde as intermediate (Larsen *et al.*, 1962). By either pathway indoleacetic acid can be produced in higher plants by invading microorganisms, and since the substance promotes the growth of the plant in exceedingly low concentrations, this reaction leads to those plant diseases that are characterised by excessive overgrowth, such as galls and nodules.

In the case of the galls produced by corn smut *(Ustilago zeae)*, it has been shown that a related non-pathogenic species does not carry out the reaction (Wolf, 1952); the indoleacetic acid, which has been isolated chemically, is present in the galls at about 20 times its normal level in the plant (Turian and Hamilton, 1960). The crown-gall organism, *Agrob tumefaciens*, produces indolealdehyde, indole-lactic acid, and other products from tryptophan; all of these arise from the breakdown of the indole-pyruvic acid first formed (Kaper and Veldstra. 1958). Some of the fungi that form mycorrhizas (swollen, inhibited, and distorted roots) convert tryptophan to indoleacetic acid very actively (Moser, 1959); mixed cultures with bacteria from the rhizosphere were even more effective.

While the reaction may occur in some normal plant tissues, it is not clear whether it is the pathway of formation of indoleacetic acid in the uninfected plant.

Desaturative Deamination

Deamination may lead to the formation of the unsaturated acid: $R\text{-}CH_2CHNH_2\text{-}COOH \longrightarrow R\text{-}CH\ CH\text{-}COOH + NH_3$

Thus, histidine is converted by organisms of the coli-typhosum group to good yields of urocanic acid, which had previously only been known from urine of dogs (Raistrick, 1917).

Histidine Urocanic Acid

The enzyme, *histidinase,* can be readily prepared from *Ps. fluorescens* (Tabor and Mehler, 1955); its *pH* optimum is 9.5. The preparation also contains an enzyme that destroys urocanic acid. Similarly aspartic acid is converted to fumaric acid by *E. coli,* this reaction.

The deamination of hydroxy-amino acids is slightly more complex. Here H_2O is removed, and the unsaturated acid that results tautom- erises to the imino acid; this is at once hydrolysed to the keto acid. L-serine yields pyruvate.

Threonine and homoserine similarly yield α -ketobutyrate. Since this compound is the first step on the route to isoleucine, threonine deaminase in *E. coli* controls the formation of this amino acid; the fact that it is inhibited by isoleucine gives a clear-cut case of "feedback control". The enzyme occurs also in *Ps. fluorescens* and has been purified from *B. cadaveris* (Williams and Mclntyre, 1955). It appears to need no co-enzyme.

The first reaction, carried out by coliform bacteria, is due to the enzyme, *serine deaminase* (Wood and Gunsalus, 1949, 1950). The extracts have been reported to need adenylic acid, zinc ions, and gluta- thione for activation.

Glutamic acid undergoes a similar reaction, but only after it has been first isomerised to the branched chain form, methyl-aspartic acid. This is then deaminated to mesaconic acid.

Barker *et al.* (1958, 1959), studying the soluble enzyme from *Cl. tetanomorphum* which catalyses this reaction, have found that a derivative of vitamin B_{12} is co-enzyme for the isomerisation step. Several B_{12} derivatives participate in similar reactions. A parallel reaction is the removal of H_2S from cysteine. The production of H_2S from peptones, presumably therefore, from cysteine, is often used as a diagnostic reaction. It is particularly active with *Proteus vulgaris,* which liberates H_2S

both from cysteine and its homologue, homocysteine, but not from thiolpropionic acid (Tarr, 1934). The amino group is therefore, needed. The cell-free enzyme preparation from Proteus, called *cysteine desulfurase*, is partially cyanide-sensitive (i.e., linked to an oxidase) but a similar enzyme preparation from *E. coli* is anaerobic (Gale, 1947). Pyridoxalphosphate acts as co-enzyme (Kallio, 1951).

Probably the initial step in both these reaction types is the formation of a hydrogen ion; it is the resulting electron shift that produces first a-amino acrylic acid and then the imino acid.

In desaturative deamination in general, it is to be noted that a second substituent in the position is needed. This may be COOH, OH, SH, or even imidazole. Desaturative deamination of simple aliphatic amino acids, at least by bacteria, has not been recorded.

Reductive Deamination

It takes place anaerobically, generally by the action of clostridia, or in a few cases through facultative organisms. Glycine is reduced to acetic acid, and both alanine and serine to propionic acid, by *Cl. putrificum* (Brasch, 1908-1909). Tryptophan is reduced by *E. coli* to £-indole propionic acid, and histidine to imidazole propionic acid (Koessler and Hanke, 1919). Aspartic acid is reduced to succinic acid by *Proteus vulgaris* and *Cl. putrificum*. Although hydrogen is evidently needed for these processes, the recorded examples are those in which amino acids alone (with inorganic salts) were the substrate. The hydrogen must come, therefore, from endogenous cell material.

In another type of reaction, the required hydrogen is supplied by the simultaneous dehydrogenation of another amino acid. Stickland (1934-1935) found that, although neither alanine nor glycine is deaminated alone by *Cl. sporogenes*, when mixed together 2 moles of glycine are reductively deaminated to acetic acid while 1 mole of alanine is "oxidised" to acetic acid, probably via pyruvic acid. The alanine yields 4H:

$$CH_3CHNH_2COOH + H_2O \longrightarrow CH_3COCOOH + NH_3 + 2(H)$$

$$CH_3COCOOH + H_2O \longrightarrow CH_3COOH + CO_2 + 2(H)$$

The hydrogen is then accepted by glycine:

$$2\ CH_2NH_2COOH + 4(H) \longrightarrow 2\ CH_3COOH + 2\ NH_3$$

Cysteine may take the place of alanine as hydrogen donor (Woods, 1936). These "Stickland reactions" suffice to maintain the growth of all of the proteolytic clostridia in a glycine-alanine medium (Nisman, 1954). They are not shown by the anaerobic micrococci, however.

These organisms convert serine to pyruvate, and threonine to α -keto- butyrate by the "desaturative" reaction, but they then liberate CO_2 from the ketoacids (Whiteley, 1957). The non-sporeforming *Diplococcus glycinophilus* can use glycine alone (Cardon and Barker, 1946, 1947); the products are acetic acid, NH_3 and CO_2; one molecule of glycine is completely "oxidised" to CO_2 and NH_3, yielding 6(H), which is accepted by three more molecules, as above.

Thus, four molecules of glycine yield three of acetic and two of CCX. If the cultures were shaken, hydrogen gas was evolved; in consequence, less hydrogen was available for reduction and correspondingly less acetic acid was formed and more CO_2 evolved. This reaction is mediated by tetra-hydro-folic acid which acts as a "one-carbon transfer system". 4). Cell-free extracts of the Diplococcus require this substance (abbreviated FH_4) as well as DPN, vitamin $B_{,,}$, and a reducible dye, in order to attack glycine (Sagers and Gunsalus, 1961). Glycine with C^{14} in the CH_3 group transfers some C^{14} to the FH_t, but C^{14} in the carboxyl group appears only as CO_2. The folic acid transfers the "1-carbon" group to another glycine molecule to form serine:

$$NH_2CH_2COOH + FH_4 + H_2O \longrightarrow FH_3\text{-}CH_2OH + CO_2 + NH_3 + 2(H)$$

$$NH_2CH_2COOH + FH_3CH_2OH \longrightarrow HOCH_2\text{-}CHNH_2COOH + FH_4$$

The serine is now converted to pyruvate and thence to acetate. This seems a complex route, but the Diplococcus does not attack other amino acids. For many other bacteria, this reaction provides the normal source of serine (Lascelles and Woods, 1954; Wright, 1955); it is thus of critical importance. In a modification of the above reactions (Stickland, 1935), proline is reduced to 8-aminovaleric acid at the expense of the dehydrogenation of either alanine or cysteine; only here, of course, the proline is not deaminated.

The addition of dithiols helps to reduce the proline, and the cell-free enzyme from Clostridia contains a dithiol dehydrogenase. DPN, Mg, and pyridoxal phosphate are co-enzymes (Stadtman, 1956). Hydroxy- proline and tryptophan can also act as hydrogen acceptors.

CL sporogenes can also deaminate arginine or ornithine alone, presumably by dehydrogenation of some molecules and hydrogenation of others, or even using O_2 as hydrogen acceptor.

It will be noticed that in the above reactions the 8-NH_2 group is not attacked. Also, the "unnatural" amino acid, alanine, which does not occur in proteins but is a constituent of the vitamin, pantothenic acid, and of co-enzyme A, is not attacked by *Cl. sporogenes.* In general, only tt-amino groups are removed in reactions of the above types. A slightly more complex reaction is involved with threonine. This is reductively deaminated with reduction also of the hydroxy group by *Cl. propionicum* (Cardon and Barker, 1947):

$$3\ CH_3CHOHCHNH_2COOH + H_2O \longrightarrow CH_3CH_2CH_2COOH + 2\ CH_3CH_2COOH + 2\ CO_2 + 3\ NH_3$$

Probably desaturative deamination first produces α-ketobutyric acid as above, and this is then partly dehydrogenated and partly reduced, somewhat as in the Stickland reaction:

$$2\ CH_3CH_2CO\text{-}COOH + 2\ H_2O \longrightarrow 2\ CH_3CH_2COOH + CO_2 + 4(H)$$

$$CH_3CH_2CO\text{-}COOH + 4(H) \longrightarrow CH_3CH_2CH_2COOH + H_2O$$

Hydrolytic Deamination and Synthesis of Arginine

Amides: When deamination occurs oxidatively and the resulting keto acid is reduced to a hydroxy acid, the impression is given that a hydrolysis has occurred, the NH_2 being converted to OH. However, it is only in the case of amides and imides that deamination by hydrolysis can occur.

The same thing happens when amides are heated with alkali. The amide group of *asparagine* is readily hydrolysed to yield aspartic acid, and that of *glutamine* to glutamic acid.

Asparaginase, the asparagine-hydrolysing enzyme, is probably widely distributed. Asparagine is an effective nitrogen source in culture media for most bacteria. An L-asparaginase has been prepared from Mycobacteria (Halpern and Grassowicz, 1957) and two separate enzymes, one for D- and one for L-asparagine, from *Brucella abortus* (Altenborn and Housewright, 1954). Glutaminase is present in several bacteria too and has been partially purified from *Cl. welchii* (Hughes and Williamson, 1952) and *E. coli* (Meister *et al.*, 1955).

Glutaminase acting in reverse is probably not the main route of glutamine synthesis, which requires ATP. However, it may have special importance because glutamine in trace amounts stimulates a number of bacterial activities. It promotes the growth and production of lactic acid by many lactobacilli and streptococci, and it promotes the formation of both toxin and alcohol by *Cl. tetani* (Lerner and Mueller, 1949; Mueller and Miller, 1949). While the mechanism of some of these effects is obscure, the action on growth may be due to a specific *transamidation* (analogous to transamination) from the amide group of glutamine to the 2-position, equivalent to a carbonyl group, of fructose-6-phosphate.

The product, 6-phosphoglucosamine, would obviously be important for cell- wall formation. Discovered in Neurospora by Leloir and Cardini (1953), the transamidase enzyme is present in *E. coli* and has been extensively purified from both (Ghosh *et al.*, 1960). Probably it is universal, and, significantly, its activity is greatest in the early stages of growth when cells are dividing rapidly (Strange and Dark, 1960). Unfortunately, at least in *Bac. cereus* and *E. coli*, there is also an enzyme that deaminates the glucosamine under the same conditions; one may guess that unless it is quickly acetylated and polymerised into cell wall, the glucosamine will not accumulate.

Guanidine and Arginine Metabolism

Guanidine (not an amino acid) is converted to urea by *Asp. niger.*

Chrzaszcz and Zakomorny (1935) found that this ability is widespread among the fungi, some accumulating urea and others further hydrolysing the urea to ammonia.

Ornithine and urea are thus formed. This reaction has been recorded for only a few micrococci and *Bac. subtilis* (Tomota, 1940-1941), but it is of particular interest because it is also the manner in which urea is produced by the animal liver (Krebs and Henseleit), the enzyme involved being termed *arginase*. There are two quite different hydrolytic breakdowns of arginine, one simple and one complex. Simple hydrolysis of the imide group yields citrulline:

$$NH{:}C(NH_2)NH(CH_2)_3CHNH_2COOH + H_2O \longrightarrow O{:}C(NH_2)NH(CH_2)_3CHNH_2COOH + NH_3 + 6000 \text{ cal}$$

This reaction, due to the enzyme arginine desimidase, is brought about by *Sc. fecalis*, *Ps. aeruginosa*, yeast, and *Cl. perfringens*, a broad enough group of organisms to suggest that the distribution is very wide, though probably irregular. Streptococci and lactobacilli, however, carry out a more complex breakdown in which CO_2 and 2 moles of NH_3 are liberated without oxygen uptake (Hills, 1940). Like so many bacterial enzymes, the system is largely adaptive, very little being formed when the cells are grown without arginine in the medium (Slade and Slamp, 1952). The non-oxidative nature of the reaction is shown by the presence of the system in *Cl. welchii* (Woods and Trim, 1942). Because the reaction results in a double hydrolysis, the enzyme was first. That the breakdown of citrulline is not a hydrolysis but a phosphorolysis and that the product is not ammonia but a new compound, carbamyl phosphate, were shown by Jones, Spector, and Lipmann (1955):

$$O{:}C(NH_2)NH(CH_2)3CHNH_2COOH + \text{-}HPO_4 \longrightarrow$$
$$NH_2(CH_2)3CHNH_2COOH + NH_2COO^- POr$$

The enzyme has been purified from *Sc. lactis* (Ravel *et al.*, 1959). The high-energy bond of carbamyl phosphate is transferred to ADP:

$$NH_2COO^- H_2PO_3 + ADP \rightarrow NH_4^+ + HCO_5^- + ATP$$

However, in *K. coli* it can be simply hydrolysed (Krebs *et al.*, 1958) to give carbamic acid, which decomposes spontaneously at *pH*. 6:

$$H_2O + NH_2COOH_2PO_3 \rightarrow H_3PO_4 + NH_2COOH \longrightarrow NH_4 + + HCOr$$

The CONH bond in citrulline must thus be one of high energy. It is suggestive that a Pseudomonas has been isolated from soil which can use citrulline as a sole source of carbon and nitrogen (Slade *et al.*, 1954); perhaps it obtains its ATP from this reaction. Some streptococci can do the same with arginine.

A still more important point is that both the citrulline-ornithine reaction and the carbamyl phosphate-ammonia reaction are reversible, so that in this way citrulline can be synthesised from ornithine, CO_2, and NH_3. These are crucial steps in the *synthesis* of arginine. In a study of the ascomycete *Neurospora crassa,* which ordinarily grows without added arginine, Srb and Horowitz (1944) found seven mutant strains which only grew if arginine were added; one of these required arginine itself, two could use either arginine or citrulline, and four could use arginine, citrulline, or ornithine. They deduced that the sequence was the same as Krebs and Henseleit had earlier worked out in the animal liver.

Precursor → Ornithine → Citrulline → Arginine

Evidently the first mutant was blocked in reaction (3), the next two at reaction (2), and the last four at reaction (1). The mutants which could not utilise ornithine were not helped by adding urea, proving that the synthesis must go by the route shown.

Using similar mutant strains, from *E. coli* (Davis, 1950), it was shown by growing the cultures side by side on a plate that the mutant requiring arginine excretes into the medium a substance that allows the mutant requiring citrulline to grow; presumably this substance is in fact citrulline. Thus, the organism cannot carry out reaction (3) and therefore, accumulates its substrate. Both strains accumulate a substance, presumably ornithine, that allows the ornithine-requiring mutant to grow.

Some of the ornithine-requiring mutants of Neurospora and of Penicillium can use glutamic acid instead, and since Neurospora can convert this acid to its semi-aldehyde, and the aldehyde by transamination to ornithine, the precursor system in Neurospora is Glutamic acid Glutamic semi-aldehyde Ornithine. Curiously enough, however, *E. coli* uses another route, since its ornithine-requiring mutants were found to accumulate not glutamic semi-aldehyde, but its acetyl derivative, as well as N-acetylornithine (Vogel, 1953).

Enzymes for the acetylation of glutamic acid, the transamination of the acetylated aldehyde, and the deacetylation of acetylornithine have since been detected. It is important too that N-acetylornithine will not allow growth of those Neurospora mutants that will grow on ornithine, although it will do so for an *E. coli* mutant.

Thus, the two pathways are quite separate. The last step in arginine synthesis has also proved to be complex. Citrulline does not simply add NH_3, since too much energy is required for this (ca. 6000 calories); instead it condenses with aspartic acid to form argininosuccinic acid.

This compound was first isolated in a study of arginine

formation by liver and is also formed from malate by dried cells of the green alga Chlorella which had been grown on urea (Walker, 1952). ATP is required in the condensation, and ADP is liberated; this satisfies the energy requirement (ATP → ADP + ca. 10,000 calories). Argininosuccinate is broken down readily by bacteria to arginine and fumarate and thus completes the route to arginine. The series of reactions centring around arginine have many ramifications in the life of bacteria. Carbamyl phosphate turns out to be an intermediate in the synthesis of pyrimidines (Lieberman and Kornberg, 1954-1955). Glutamic semi-aldehyde is the precursor of proline as well as ornithine (above). The CO_2 given off in arginine breakdown acts as a growth factor for certain streptococci that need CO_2 but cannot produce it from sugar (Gale, 1945b); these can grow on ornithine instead of arginine, but only if another source of CO_2 is supplied. Finally, the synthesis of the enzyme interconverting ornithine and citrulline in *E. coli* is suppressed by arginine, so that added arginine inhibits its own synthesis; this is the phenomenon of "feedback control".

Urea Hydrolysis

This important hydrolytic reaction:

$$NH_2CONH_2 + 2\ H_2O \rightarrow (NH_4)_2CO_3$$

is brought about by a limited number of bacteria. If urea be mixed with peptone and inoculated with soil, the predominating organisms are sporeformers, called by Beijerinck (1901) Urobacillus but now considered as *Bae. pasteurii* and its congeners. These organisms, as well as a non-sporeformer isolated by Sohngen (1909) as *Urobac. jakschii*, can hydralyse urea in concentrations up to 10 per cent. They grow best on alkaline media (up to *pH* 11) and can do without urea on media made to pH, 9 and containing 1 per cent NH_4Cl (Gibson, 1934). *Bac. pasteurii* has an actual requirement for ammonia in addition to peptone, although other urea-hydrolysing bacilli do not (Born- side and Kallio, 1956). Beijerinck also isolated a Sarcina *(Sporosarcina ureae)* with similar physiology, which is most unusual because it is motile and forms spores, yet otherwise is a typical-looking Sarcina. It is evidently widely distributed in soils (Gibson, 1934).

If, instead of peptone, calcium malate be used in the enrichment culture with 2 per cent urea, non-sporoformers develop that can hy- drolyse urea only at concentrations up to 3 per rent (Sohngen, 1909).

These include *M. ureae* (often also found in old urine) and a red organism, "*B. erythro genes*" with a light-sensitive pigment. *M. aureus*, as well as certain yeasts, including Cryptococcus and Rhodotorula, also hydrolyses dilute urea solutions, and the reaction is reliable enough to help in identifying them (Seeliger, 1956).

None of these organisms can grow in the absence of a carbon source in addition to urea. This is partly because the urea-hydrolysing bacilli require traces of biotin, thiamine, or nicotinic acid (Bornside and Kallio, 1956) but partly because the energy liberated by urea hydrolysis is small. However, the reaction is so vigorous that very little carbon is needed; 20 mg of asparagine allow hydrolysis of 3600 mg of urea by *Urobac. jakschii*, and *M. ureae* decomposes up to 2000 times its wet weight of urea per hour (Burchard, 1899).

The hydrolysis is due to the enzyme *urease*, which is also present in some leguminous plants, notably the jack bean and soybean, but the urease from *Bac. pasteurii* is much more active than the jack-bean enzyme (Larsen and Kallio, 1954).

Its Qm, as would be expected for a hydrolysis, is only about 1.5. This enzyme is one of many whose activity depends on the presence of sulphhydryl groups. It is strongly inhibited by iodoacetate, arsenite, and copper and mercury salts, as well as organic mercury compounds; these react with the group. In *M. aureus* the sensitivity of urease to inhibitors is limited to organisms grown in the presence of urea in the medium (Yall and Green, 1952). But the growth and urea hydrolysis by Urobacillus can be dramatically inhibited by traces of these reagents. The importance of urea hydrolysis lies, of course, in the fact that it is a typical animal excretion product, and that large quantities of it are constantly being added to soil arid to natural waters.

Creatine Breakdown

The bacterial breakdown of creatine (N-methyl-guanidine acetic acid) differs from simple deamination in that it involves splitting off of urea; in this it has much in common with the breakdown of arginine. Creatine is attacked oxidatively by only a few bacteria. In the case of a Pseudomonas resembling *Ps. aeruginosa,* isolated from urine, the breakdown is rapid, leading to urea and ammonia (Kopper and Beard, 1947). The anhydride of creatine, creatinine, is brought into equilibrium with creatine by an enzyme called creatinomutase.

This equilibrium is brought about by a soil organism obtained from enrichment cultures on a creatinine medium. After growth on this medium the bacterium develops an enzyme system converting creatine to urea and sarcosine (Akamatsu and Miyashita, 1952). The sarcosine is demethylated to glycine, and the glycine liberates ammonia, giving the following sequence.

The first stage exactly parallels the action of arginase, and an enzyme for it has been prepared from a soil bacterium that did not attack creatinine (Nimmo-Smith, 1949).

Reactions other than Deamination

The breakdown of amino acids is complicated by a number of side reactions that do not yield ammonia and therefore, reduce the rate of the nitrogen cycle in nature.

Assimilation

In the first place, synthesis of bacterial or fungal protoplasm absorbs amino acids. Those organisms, like *E. coli* or *Asp. niger,* that are able to grow on a single amino acid or on a carbohydrate plus ammonium salts, must have the power of synthesising all ordinary amino acids from NH_3, like green plants. It usually requires oxygen and, in any event, requires some source of energy. If a single amino acid is the sole nutrient, there will in general be an excess either of carbon compounds or of nitrogen; in the latter case ammonia will be excreted into the solution. Alanine seems to give nearly the right balance of C and N, and can be assimilated very efficiently by fungi like Leptomitus (Schade and Thimann, 1940). On the other hand, glutamate and asparagine, when oxidised by *Brucella abortus*, give a partial yield of NH_3, and CO. It is evident that part of the amino acid is assimilated while part is being oxidised. Glycine is usually a very poor growth substrate, and few bacteria, protozoa, or fungi will grow on it alone. *Diplococcus glycinophilus* is outstanding in this respect. The polyglutamate capsule represents another drain on the amino acid supply.

Pigment Formation

In the second place, bacteria form a number of pigments in which nitrogen becomes temporarily immobilised. One of these is the group of black compounds called *melanins*, which are formed from tyrosine by the enzyme polyphenol oxidase. In this process, the tyrosine is first oxidised to dihydroxyphenyl alanine or *dopa;* Dopa, being a dihydric phenol, is further oxidised to the orthoquinone, and this undergoes an intramolecular hydrogen transfer to form an indole ring. A second oxidation yields the quinonoid structure called *hallachrome,* a red pigment, and this undergoes hydrogen transfer to form 5,6-dihydroxy-2-indole carboxylic acid. The last-known stage is that this loses CO_2 and condenses to the black melanins.

The series of reactions, elucidated by Raper (1928), and confirmed by several workers (Beer *et al.*, 1949, and Clemo *et al.*, 1950, 1952), are as follows:

> Later C^{14} experiments show, however, that melanin formation is more complex than simple condensation, because about half of the CO_2

while another mutant forms a different pyrrole compound, 2-aldehydo, 3-methoxy,5-pyrryl. When cultures of this second white mutant are exposed to vapours of the methylamyl-pyrrole, they turn red immediately, i.e., the two pyrrole compounds condense into prodigiosin (Wasserman *et al.*, 1960): This may give a simple explanation for some of the complex genetical findings with crosses between Serratia strains. It also explains Santer and Vogel's observation (1956) that with two white mutants growing side by side one produces a diffusible precursor from which the other forms red pigment. There may be several related pigments in the Serratia gene-complex, because in some strains (e.g., var. *Kiliensis)* the pigment diffuses out into the medium, while in others it is wholly intracellular. In some, the red compound is accompanied by a blue pigment of apparently similar structure, but a molecular weight about 775, i.e., perhaps a double molecule (Green *et al.*, 1956). Another triple-ring pigment is the beautiful violet compound, violacein, formed by *Chromob.* (= Agrob.) *violaceum*. The elucidation of this structure has proved very difficult; after several attempts by Marchal in France and Beer in England, it was finally shown by Ballantine *et al.* (1958) to be IV:

Thus, it contains 5-hydroxyindole (A), pyrrolenone (B), and oxindole (C). A marine strain isolated from codfish yielded the same pigment. Presence of the two indole rings suggests that violacein is an oxidation product of tryptophan. This agrees with the fact that a suspension of *C. violaceum* actively oxidises tryptophan to the 5-hydroxy derivative and also breaks it down to indoleacetic and 5-hydroxyindoleacetic acids (Mitoma *et al.*, 1956). Indeed, C. *violaceum*, suspended in a solution of L-tryptophan alone, can form the pigment (DeMoss and Evans, 1959). An oxindole compound is known as an oxidation product of indoleacetic acid by the fungus Omphalia (Stowe *et al.*, 1957), so that this too can be formed from tryptophan. Indole is probably an intermediate, and when the cells are lyophilised they convert indole to indigo instead of to violacein (Sebek and Jager, 1962).

Many workers have wondered what function these (and other) pigments serve. The quantity is of course very small usually only a fraction of 1 per cent of the dry weight. But at this concentration other pigments such as hemins or flavoproteins can be critically important in the life of the cell. However, colourless mutants, e.g., those of Serratia, seem at no disadvantage as regards growth rate or nutrition. Pyocyanin can be reversibly oxidised and reduced and can function as a coupler between some dehydrogenases. Prodigiosin is a fairly effective antibiotic, inhibiting the growth of *M. aureus* at a concentration *as* low *as* 1:30,000 (Efimenko *et al.*, 1956). But whether these functions are served in nature remains obscure.

Decarboxylation

This is the most important side reaction. It is widely carried out by members of the coliform group, clostridia, and lactobacilli (Rodwell, 1953), as well as certain fungi, and is comparable to decarboxylation of the keto acids.

Amino acid decarboxylations are of great biochemical importance. The resulting amines were isolated by early workers from putrefying meat; they have unpleasant odours, are basic, and are toxic to animals. Brieger (1886) termed these bases *ptomaines* (the word was invented by Selmi in 1878) and considered them responsible for the "ptomaine poisoning" which follows the ingestion of bacterially spoiled meat or fish. However, although such bases are indeed toxic on injection (Berthelot, 1918), yet when fed by mouth, ordinarily spoiled meat has little ill effect (Savage, 1920).

For the concentration of the bases to reach anywhere near the toxic level, the meat must be so far gone that it would be completely unpalatable. Further, these bases are present in the normal alimentary canal as a result of bacterial action, so that they cannot be highly toxic.

Examples of the decarboxylation of amino acids are:

1. *Ornithine* produces *putrescine,* isolated by Ellinger (1899, 1900), from a mixed culture and subsequently by Gale (1941) from *Cl. septicum:*

   $$NH_2(CH_2)_3CHNH_2COOH \longrightarrow NH_2(CH_2)_3CH_2NH_2 + CO_2$$

   The same base was obtained from arginine, doubtless by way of ornithine (Ackermann, 1910a). True food poisoning is either due to production of a very active toxin by the anaerobe *Cl. botulinum* or to actual bacterial invasion of the intestine by organisms present in the meat, such as *Salmonella typhimurium.* The noticeable extent of putrefaction in meat has little relation to its toxicity. Thus, many outbreaks of "botulism" have arisen from the eating of quite unsuspected food on which *Cl. botulinum* had grown and produced toxin and died out again, without causing much obvious breakdown. The botulinus toxin is a protein (Abrams *et al.,* 1946) and is not heat-stable, so that 10 minutes' boiling of infected food will render it perfectly safe.

2. *Arginine* is converted to *agmatine* by intact *E. coli* (Gale 1940), or by the enzyme from it (Umbreit and Gunsalus, 1945):

   $$HN{:}CNH(CH_2)_3CHNH_2COOH \rightarrow HN{:}CNH(CH_2)_3CH_2NH_2 + CO_2$$

3. *Lysine* produces *cadaverine:*

   $$NH_2(CH_2)_4CHNH_2COOH \longrightarrow NH_2(CH_2)_4CH_2NH_2 + CO_2$$

Cadaverine is so named because of its odour of corpses, in which it doubtless arises by bacterial action. Both putrescine and cadaverine are produced by *Aspergillus oryzae* from soybean proteins, so that they are by no means products only of bacteria (Yamada and Ishida, 1926).

4. *Glutamic add yields y-amino butyric acid:* This reaction occurs with *Rhizobium leguminosarwn* (Virtanen *et al.*, 1938), *E. coli* (Gale, 1940; King and Fletcher, 1950), and *Cl. welchii* (Gale, 1945). The Aerobacter species differ from Escherichia in being unable to decarboxylate glutamic acid. *Leucine* and *valine* are decarboxylated by *Proteus* (Haughton and King, 1961).

5. *Aspartic acid* yields */3-alanine:*

   $$HOOOCH_2CHNH_2COOH \longrightarrow HOOOCH_2CH_2NH_2 + CO_2$$

   This reaction, exactly comparable to (4) above, is carried out vigorously by cultures of Rhizobium (Virtanen *et al.*, 1938), and also by *Cl. welchii* in the presence of pyruvate (Meister *et al.*, 1951). Very few other bacteria can bring it about.

6. *Tyrosine* yields *tyramine,* which has interest for its marked pharmacological action on animals: It was first obtained with an intestinal organism (Berthelot and Bertrand, 1912), and with *Proteus vulgaris* and *E. coli* growing on tyrosine solution (Sasaki, 1914). More recently Gale and co-workers (1940, 1946) found that 500 strains of *Sc. fecalis,* as well as some other streptococci (but not Sc. *lactis),* carry out the reaction. It is less common among coliform bacteria, being found in less than 10 per cent of the strains investigated (Stadler and Neus, 1935; Gale, 1946).

7. *Tryptophan* (indolealanine) has been reported to yield *trypta-mine:*

   $$ICH_2CHNH_2COOH \longrightarrow ICH_2CH_2NH_2 + CO_2$$

   (I = indole nucleus)

No organism capable of doing this has been found since Berthelot and Bertrand's report in 1912, in spite of examination of over 1000 cultures by Gale and co-workers. The corresponding decarboxylation of 5-hydroxytryptophan to 5-hydroxytryptamine, or serotonin, however, goes readily in the presence of a soluble enzyme in the kidney (Udenfriend, 1959) and is important in animal physiology because of the potent neurological action of serotonin. However, this reaction is not yet known in bacteria.

8. *Histidine* yields *histamine,* which powerfully contracts the animal uterus and also plays a part in surgical shock and in allergic reactions. Unlike tyramine, it lowers the blood pressure. The decarboxylation played an interesting part in the discovery of histamine. Barger and Dale (1910), studying the toxic action of ergot (from rye grains parasitised by the Ascomycete *Claviceps purpureum),* found that neither of the two substances, ergotine and tyramine, known to be present in ergot, was responsible for the uterine contraction that is typical of ergot poisoning. They therefore, looked for another, unknown substance in the ergot extract and arrived at the conclusion that it must have arisen during bacterial action on their stored ergot extracts. On allowing their preparation (or later, meat extract and yeast extract) to putrefy, a new base, apparently related to histidine, was isolated from the mixture. At the same time, Ackermann (1910) had obtained and identified histamine from the partially anaerobic putrefaction of histidine in presence of glucose, peptone, arid calcium carbonate. Comparison of their compound with a sample of Ackermann's proved it to be histamine. The organism of Berthelot and Bertrand (above) also produces histamine, as do strains of *S. typhosa* (Mellanby and Twort, 1912) and *E. coli* (Koessler and Hanke, 1919). Berthelot (1918) showed that if proteolytic Clostridia were grown in

blood, they set free histidine from the blood protein as expected; if now a decarboxylating organism of the Aerobacter type were introduced, histamine was rapidly formed at 37° C. Thus, infection of wounds by a proteolytic and a decarboxylating organism together could possibly give rise to sufficient histamine to have serious physiological results.

9. *Diaminopimelic acid* (not found in proteins) yields *lysine.* This is important because it is a normal path of lysine formation; in other words this decarboxylation differs from the others in that its product is a normal amino acid. Diaminopimelic acid occurs in the cell walls of sporeformers and is dehydrogenated to dipicolinic acid during spore formation. In the vegetative cell, however, it is decarboxylated (Powell and Strange, 1957). The de- carboxylase is also present in small amount in the spores themselves.

10. *Threonine* yields *aminoaceione:*

$$CH_3CHOHCHNH_2COOH \rightarrow CH_3COCH_2NH_2 + CO_2 + 2(H)$$

This is an oxidative decarboxylation, and α -amino-isopropanol is not an intermediate (Elliott, 1960); the dehydrogenation thus takes place first. The enzymes concerned are *decarboxylases.* In *E. coli,* they are produced at *pH* 5, but not at *pH. 7* (Gale, 1940); presence of the amino acid in the medium generally favours production of its de- carboxylating enzyme. Presence of sugar in the medium also favours decarboxylation. The pH optimum for the action of these enzymes is from 4 to 5. The reaction is completely inhibited by cyanide at about 3 X 10^{-5} M, as well as by silver and mercury ions (Taylor and Gale, 1945), and semi-carbazide (Meister *et al.*, 1951).

The decarboxylases are so specific that they can be used to determine certain amino acids precisely. The procedure is to add a washed suspension of the bacteria, or ari acetone-dried powder, to the amino acids in a manometer and estimate the CO_2 evolved. Only a few milligrams of amino acid are needed.

Since the enzymes act at acid pH, the CO_2 is not absorbed and the determination is complete in about 30 minutes. Only L-amino acids are attacked (i.e., natural forms), but hydroxy-amino acids are attacked along with their non-hydroxy counterparts.

The organisms are grown in a protein-hydrolysate medium containing 1 to 2 per cent glucose, since glucose promotes formation of decarboxylases. The following determinations (Gale, 1943-1946; Krebs, 1948; Meister *el al.*, 1951) are quantitative. In the case of tyrosine decarboxylase of *Sc. fecalis*, the purified enzyme was found to require a co-enzyme. The latter is pyridoxal-5- phosphate, or vitamin B_{12} phosphate (Gunsalus and Bellamy, 1944; Umbreit *el al.*, 1945). The same compound is co-enzyme for all other decarboxylations studied, as well as for transaminations.

The structure below was confirmed by synthesis (Heyl and Harris, 1951) and demonstration that the 3-phosphate is inactive. The highest A mixture of pyridoxal with excess of ATP will also act as co-enzyme, although with greatly reduced efficiency. Evidently, therefore, phosphorylation is carried out by ATP. The enzyme for this, *phosphokinase,* has been detected in several organisms and purified from yeast (Hurwitz, 1952); it is not specific for pyridoxal but will also use ATP to phosphorylate pyridoxine and pyridoxamine.

It can also phosphorylate other analogues, some of which inhibit decarboxylation by preventing the union of co-enzyme with apoenzyme. The apoenzyme, which is, of course, a protein, can be extracted from *Sc. fecalis* which has been grown without vitamin B_{12}; when pyridoxal phosphate is added to it, it will decarboxylate tyrosine (Umbreit *et al.*, 1945), and it can therefore, be used to determine pyridoxal phosphate in tissues (Sloane-Stanley, 1949). The elucidation of the mode of action of pyridoxal phosphate in decarboxylation was bound up with the study of its action in trans- amination. The finding that both the aldehyde (pyridoxal) and the amine (pyridoxamine) occur naturally led Snell (1944) to suggest that these two forms might be convertible into one another and thus catalyse transamination to other aldehydes and ketones. This was supported by evidence that transamination in animal tissues, as well as tyrosine decarboxylation in *Sc. fecalis*, was impaired in vitamin B_6 deficiency. Both could be restored by adding pyridoxal phosphate or pyridoxal plus ATP.

TABLE: Activity of Amino Acid Decarboxylaaes in* E. coll *Grown on Different Media

(Data of Gale, 1946)

(Activity expressed as Qco_2 at 30° C and the optimum pH)

Amino Acid Added To The Medium

Amino Acid as Substrate in Test	*None Ornithine Glutamatc ITistidine Tyrosine*	*All*	*Lysine*	*Arginino*
L(H-)Lysine	4	104	210	4
L(-f)Arginine	0	330		27
L(+)Ornithine	3	145		225
L(+)Olutamate	45	100		88
L(?)Histidine	0	18		7
L(?)Tyrosine	0	63		60

Since, during transamination, the optical activities of the amino acids are not changed, they must be attached rigidly to the protein at not less than three points. The formulae above are oversimplified, since Ca^{++} or another divalent metal plays an important part; indeed, some reactions can occur without the protein (i.e., non-enzymatically) if a metal ion is present.

As mentioned above, presence of the amino acid in the growth medium *induces* formation of its decarboxylase.

Indole Production

The production of indole and skatole from tryptophan is a change of another type:

> Indole was first detected by W. Kuhne in 1875 among the products of proteolysis and isolated by Nencki in the same year from protein incubated three days at 40° C; its constant occurrence as a bacterial product led Hopkins and Cole (1903) to search for its precursor and hence to discover tryptophan, a constituent of nearly all proteins.

Indole formation is characteristic of *E.* coli, *Proteus vulgaris,* and many related organisms. It is readily detected because of the numerous sensitive colour reactions given by indole and its derivatives, especially with dimethylaminobenzaldehyde in HCl; these have made indole formation a routine diagnostic test.

We have seen above that oxidative deamination of tryptophan by many bacteria and fungi may yield indoleacetic acid. One might, therefore, suppose that further breakdown of this acid would yield indole. Bacterial action on indoleacetic acid, however, does not yield indole. (It probably yields skatole [3-methyl indole] by decarboxylation, although this has never been shown with pure cultures). None of the expected intermediates indoleacetic, indole-pyruvic or indole-dicarboxylic acids, indole-ethylarnine, indole-aldehyde or skatole yields indole with *E. coli* under the conditions in which it is formed from tryptophan (Frieber, 1922; Woods, 1935). The only precursor other than tryptophan that Woods could find to produce indole was a mixture of indole-pyruvic acid and ammonia; since oxidative deamination can be reversed the organism probably synthesised tryptophan from these constituents. In other words, tryptophan itself is required for the production of indole.

The failure of possible intermediates to produce any indole led to the idea that tryptophan is split directly into indole and alanine, but preparation of a cell-free in dole-forming enzyme from autolysed dried *E. coli* has shown that alanine is not produced (Wood *et al.*, 1947; Dawes and Happold, 1949). This enzyme is called *tryptophanase,* and requires pyridoxal phosphate as co-enzyme. The combination converts tryptophan into indole, pyruvate, and NH_3, in ratios almost exactly 1:1:1. No oxygen is required. It does not deaminate alanine or serine, which indicates that alanine is not an intermediate product, i.e., the deamination is simultaneous with the reductive separation of the side- chain:

> The mechanism for this curiously smooth rupture of a carbon-to- carbon bond has been much discussed. It is catalysed by traces of Cu and Co in alkaline medium. The most reasonable mechanism suggested involves preliminary isomerisation to pseudotryptophan, formation of an anion by the dehydrogenation, and subsequent migration of the electrons.

The liberated a-amino acrylic acid in the tautomeric imino-form would be at once converted to pyruvic acid. Intact cells would, of course, oxidise the pyruvate away very rapidly, which accounts for its not appearing in cultures. The cell-free enzyme after purification does not destroy pyruvate. Tryptophanase is produced in *E. coli* only in presence of tryptophan, i.e., its formation is induced. When used as an antigen in rabbits, the resulting antisera act as tryptophanase inhibitors; since they appear to be specific for different bacterial species (Dolby and Happold, 1951), a family of closely related enzymes may be involved. Like the amino acid oxidases, histidinase, and other enzymes, tryptophanase formation is prevented by glucose, M/10 glucose in the medium suppressing its production 100 per cent (Freundlich and Lichstein, 1960). The enzyme that catalyses the reverse reaction, synthesising tryptophan from indole and serine or from indoleglycerol phosphate, reacts to glucose in the opposite way, its formation being promoted (Monod and Cohen-Bazire, 1953).

The production of phenol from tyrosine by "*B. coli phenologenes*" (a strain of *E. coli*) has also been reported (Rhein, 1918). As with indole, phenol is not formed from any of the probable intermediates of deamination, either oxidative or reductive:

> Doubtless the reaction will turn out to be of a type similar to that with tryptophan.

Indole production by *E. coli* in the intestines, if sufficiently vigorous, may occasionally lead to the excretion in the urine of indoxyl, an oxidation product of indole, as a glucoside or sulphate known as indican. In such urine, on standing, the indican is again hydrolysed to indoxyl, two molecules of which readily oxidise to indigo blue: the urine thus deposits a deep blue precipitate, and the condition is known as indicanuria:

> The alanine deaminase or dehydrogenase normally operates in the direction of synthesis and is thus a major route of ammonia assimilation in the bacilli. Glutamic dehydrogenase fulfils the same function for most other bacteria and many fungi. Alanine and ghitamic acid, once formed, readily take part in transamination reactions with a-keto acids. For most other amino acids, therefore, the route of synthesis is merely the route to the corresponding a-keto acid, and the last step is transamination (proline, threonine, serine, and tryptophan are exceptions). For completeness, the known reaction sequences leading to the appropriate keto acids in bacteria will be reviewed here; some of them are treated elsewhere.

Pyruvate is formed directly from sugar by aerobic and anaerobic routes. *Oxalacetic* and *a-keto-glutaric acids* similarly [illegible] the Krebs cycle. *α-ketoisovaleric acid* ("ketovaline") is formed from acetolactate. Mutants of *E. coli*, requiring valine, accumulate α-acetolactate; the cell-free extracts will form acetolactate from pyruvate, and normal cells fed C^{14}-acetolactate produce C^{14}-valine in the protein (Umbarger and Brown, 1958). Acetolactate undergoes rearrangement and reduction to dihydroxy-isovalerate:

The dehydrogenase for reaction using TPN (= NADP), probably also causes, or favours, reaction 2. The fourth is due to a dehydrase, present in extracts of normal *E. coli* and absent in certain valine-requiring mutants, which accumulate the dihydroxy-acid. The final step is transamination with glutamic acid. α *-keto-ft-methylvaleric acid* ("keto-isoleucine") is formed by parallel reactions, the lower CH_3 group being replaced by C_2H_5. Instead of forming acetolactate, pyruvate condenses with α-ketobutyrate.

The rearrangement, reduction, and dehydration, at least by E. *coli* mutants, are then exactly as for valine. Furthermore the enzymes are actually the same for each step, the condensing enzyme, the isomerase, dehydrogenase, and dehydrase operating equally on the two members of each pair of homologs. This is shown not only by biochemical similarities between the enzymes, but also genetically by the fact that many apparently "single-step" mutations cause *E. coli* or *S. typhimurium* to require *both* isoleucine and valine for growth (Umbarger *et al.*, 1960-1961). The fact that the condensing enzyme requires Mg and diphosphothiamine indicates its close relationship to carboxylase, which can cause not only decarboxylation of pyruvic acid to acetaldehyde but also its condensation to acetoin.

α *-ketoisocaproic acid* ("ketoleucine") is derived from "ketovaline" by a further condensation with acetyl-CoA.

This is isomerised to the a-hydroxy isomer and dehydrogenated to the α-keto acid. The final step is again transamination with glutamic acid. Thus these three α-keto acids, and therefore, the aminoacids valine, isoleucine and leucine, arise on the same pathway: Now, formation of bacterial enzymes is frequently repressed by presence of their products. In this case the presence of excess of *either one.*, e.g., valine, would prevent the formation of enzymes synthesising *all three* amino acids. Perhaps to offset this complication, £. *coli* and *S. typhimurium* show the peculiar effect that enzymes for the early steps, i.e., threonine deaminase and the dehydrase, are *not* repressed by isoleucine or valine alone, but only when leucine also is added. By contrast, formation of the decarboxylase, which makes only leucine, *is* repressed by leucine alone (Freundlich *et al.*, 1962).

Phenyl pyruvic and *p-hydroxyphenylpyruvic* acids, precursors of phenylalanine and tyrosine, are formed from sugar via sedoheptulose diphosphate and a 7-carbon deoxy-keto-heptose which undergoes cyclisation. The product is gradually dehydrogenated to the aromatic state and takes on a pyruvic side chain to form the immediate precursor, prephenic acid. This loses H_2O and CO_2 directly to form phenylpyruvic acid, but just how the p-hydroxy group is introduced for tyrosine is still not clear; it may require an oxygenation. In the formation of *histidine*, the expected keto acid is replaced by a keto-phosphate, imidazole-acetol phosphate. After this has been trans- aminated (by glutamic acid), the phosphate is hydrolysed off and the resulting alcohol oxidised to an acid.

Serine is formed by transformylation from glycine, the CH_2OH group being carried over by tetrahydrofolic acid (FH_4). The formyl group can come from formic acid (hence from CO_2), from glyoxylic acid or other sources, to produce formyl-tetrahydrofolic acid, which is then reduced (F = folic acid):

$$HCOOH + FH_4 \rightarrow FH_3\text{-}CHO + H_2O$$

$$FH_3\text{-}CHO + DPNH + H^+ \rightarrow FH_2\text{-}CH_2 + H_2O + DPN^+$$

$$FH_2\text{-}CH_2 + NH_2CH_2COOH(B_6) + H_2O \rightarrow HOCH_2\text{-}CHCOOH + FH_4 + (B_{12})\ NH$$

In the last step, catalysed by *serine hydroxymethylase*, the glycine is combined with pyridoxal phosphate (vitamin B_{12}), accepts the $HOCH_2$ group, and splits off the B_{12} again. The enzyme system isolated from *E. coli* has a vitamin B_{12} derivative as its prosthetic group and requires ATP, DPNH, and FAD as well as the folic acid and B_6 (Hatch *et al.*, 1961). The intermediate FH_2CH_2 (N^5-N^{10}-methylene-tetrahydrofolic acid) is a general donor of methylene groups; it methylates uracil to thymine and ethanolamine to choline also (Kisliuk and Sakami, 1955). Hence this family of reactions is of central importance. *Threonine* is formed from aspartate by the following reactions:

ATP TPNH TPNH

Aspartic → Aspartyl phosphate → Aspartic semi-aldehyde → ATP

Homoserine → O-phosphohomoserine → Phosphate and threonine.

This route appears the same in Neurospora, yeast, and *E. coli*. Again it offers an example of "feedback control," the first reaction being inhibited by threonine. The last reaction, due to threonine synthetase (purified from Neurospora), uses pyridoxal phosphate (B_6) as co-enzyme; the combination allows electron migration and hence the uptake of OH at the O-position (Flavin and Slaughter, 1960).

A Threonine-requiring Mutant did not Possess this Enzyme: The pyrroline ring is formed from glutamic semi-aldehyde by loss of ELO and is then reduced to proline. Hydroxylation to *hydroxy- proline* occurs (at least in animal tissue) only on proline bound in protein. *Methionine* is formed from two amino acids, cysteine and homo- serine. Homoserine is not present in proteins, but occurs free in small amounts, especially in seedlings; some mutant fungi require it for growth. In Neurospora, a number of strains which cannot utilise sulphate require sulphur-containing amino acids for growth. Some of these can use only methionine or homocysteine, $HS(CH_2)_2CHNH_2$,COOH. One requires methionine itself, and this one excretes into the medium a new compound, cystathionine.

Cysteine Homoserine Cystathionine Homocysteine

It is readily converted to homocysteine. The methylation of homocysteine is carried out by *E. coli* with serine as CH_4

source (Szulmajster and Woods, 1960). The enzyme requires a folic acid derivative as co-factor and forms glycine from serine under the same conditions.

Breakdown of Amines

Reasoning that many anaerobic processes lead to formation of compounds relatively resistant to biological decomposition, de Jong (1926) set out to study the organisms responsible for the breakdown of higher alcohols, fatty acids, and amines. While none of the known bacteria tested would grow on the amines as carbon and nitrogen source, organisms could be isolated from natural waters using enrichment cultures of these compounds. These included a group of related Pseudomonads, named *Ps. aminovorans* (four types), two new species of Mycobacterium, and some non-motile gram-negative forms which could not be placed in any recognised group. Since these last are characterised by ability to grow on simple amines, de Jong named them *Protaminobacter*. All these bacteria are aerobic and catalase-positive, so that the attack is oxidative in all cases. Since the growth was the characteristic studied, the nature of the intermediate products, if any, in the oxidation of these compounds remains unknown.

The di- and tri-sub- stituted amines are evidently less attackable than the monoamines. On the higher dialkylamines almost the only organism able to grow is *Protaminob. alboflavum*. The Mycobacteria are clearly limited to monosubstituted amines. The amines formed by decarboxylation of amino acids are somewhat more readily attacked; tyramine is oxidised at least partially by *Ps. pyocyanea* and also by some strains of *E. coli* and its relatives, but only when the organisms were previously grown on a medium containing some tyramine. The same is true for the oxidation of histamine by *Mycob. smegmatis* (Roulet and Zeller, 1945).

Although not much is known about the monoamine oxidases of bacteria, the enzymes attacking diamines have been well studied. Pentamethylene diamine (cadaverine) and histamine are attacked by the two Mycobacteria, as well as by *Mycob. smegmatis*. *Mycob. lacticola* oxidises histamine and putrescine (Franke and Schillinger, 1944), and other Mycobacteria, including those of tuberculosis in man and birds, attack putrescine. *Pseudomonas pyocyanea* oxidises putrescine, cadaverine, and agmatine (Gale, 1942); *Proteus morganii* oxidises putrescine and cadaverine and *M. aureus* oxidises cadaverine. These enzymes seem to be always present irrespective of growth medium, but they can be increased by culturing in presence of the base concerned.

Roulet and Zeller (1945) observed that when two diamines were given simultaneously to *Mycob. smegmatis,* the oxygen uptake was never greater than on one alone, and also that if one of the two diamines was oxidised very slowly and the other rapidly, then the resulting combined oxygen uptake was intermediate between the values found with the diamines separately. For example, the oxygen consumption (corrected for that of blanks without amine) was as follows: putrescine 920; cadaverine 170; putrescine + cadaverine 625. (Expt. 2): putrescine 1005; histamine 5; putrescine + histamine 725.

Similar observations were made with *E. coli* (Zeller, 1951). It is, therefore, concluded that all the oxidations are caused by one enzyme, *diamine oxidase.* Besides the diamines, this enzyme attacks histamine but not other monoamines. Even with diamines a separation of the two NH_2 groups by 14 or more carbon atoms prevents them from being attacked by the diamine oxidase. The marked differences between the attacking abilities of the organisms indicate that for the monoamines several enzymes must be involved. Since the optimum *pH* for the oxidations is 7.9 to 9.5, they will not proceed at the *pH* (4.5 to 6) at which the amines were formed (Gale 1942). As a rule, too, the bacteria that form the amines will not oxidise them. The oxidations are inhibited by some antibiotics, including streptomycin, dihydrostreptomycin, and chloramphenicol (Owen *et al.,* 1951), but the significance of this is uncertain since they are also inhibited by numerous synthetic amines and the pigment pyocyanine. It is important that the oxidations are inhibited by semi-carbazide, dimedon, and hydroxylamine, since all these are reagents combining with aldehydes (Roulet and Zeller, 1945). This indicates that the first product is the aldehyde, and indeed Gale (1942) was able to isolate hydroxyphenylacetaldehyde when *Ps. pyocyanea* attacked tyramine. A diamine would correspondingly yield an aminoaldehyde.

Since the oxidation is accompanied by formation of H_2O_2, which may be secondarily destroyed by catalase, the reduced diamine oxidase (DOH_2) is evidently reoxidised by a flavoprotein which in turn forms H_2O_2 when it reacts with oxygen. With the enzyme which has been purified from animal tissue (Tabor, 1951), the aminoaldehyde can be further oxidised to the acid, either by oxygen, using aldehyde oxidase, or with DPN as hydrogen acceptor.

In the early stages of the reaction less than 1 atom of oxygen is used per molecule of NH_3 formed (Fouts, 1953), because the aminoaldehyde can itself act as hydrogen acceptor for the reduced enzyme and thus be converted to an aminoalcohol. The whole system can be represented thus (diamine oxidase = DO): Formation of the imine may also occur in the oxidation of mono- amines and would explain why di-substituted amines are so seldom attacked.

Breakdown of Nucleic Acids and of their Components

Nucleic Acids, as Mentioned many Times Earlier, are of Two Types: RNA containing ribose with the bases adenine, guanine, cytosine, and uracil, and DNA containing 2-deoxyribose with adenine, guanine, cytosine, and thyrnine. In the first stages of their breakdown, i.e., depolymerisation and hydrolysis to the nucleotides (base-ribose-5- phosphate), a number of enzymes participate; these often differ from one species to another. Active RNA-ases are excreted by *Bac. subtilis, megaterium*, and *mesentericus* (MacFadyen, 1934), and by Aspergillus and other fungi (Otani, 1935). A DNA-ase has been prepared from haemolytic streptococci, which excrete it into the medium (McCarty, 1949). It requires Mg and is inhibited by RNA. A similar enzyme extracted from *E. coli* acts only on denatured or 1 -stranded DNA (Lehman, 1960).

In the second stage of breakdown the hydrolysis of phosphate is due to widely distributed phosphatases and pyrophosphatases.

The ribose can be hydrolysed off too, or it may be removed by *phosphorolysis* as ribose- 1 -phosphate, e.g., *E. coli* converts uracil and cytosine ribosides, with inorganic phosphate, to the free bases and ribosephosphate (Kal- ckar, 1947). *Lactobacillus pentosus* actively hydrolyses purine and pyrimidine ribosides, probably by a pair of related enzymes (Wang, 1955), and similar enzymes for uridine and for purine ribosides occur in yeast. The hydrolysis may be associated with deaminatiori, as when *E. coli* and Aerobacter attack adenylic acid to give H_3PO_4, ribose, and NH_3, the adenine being deaminated to hypoxanthine (Stephenson arid Trim, 1938).

A similar reaction is carried out by an enzyme from *Asp. oryzae*, or from the Taka-diastase made from its mycelium, as well as from *Neurospora crassa* (Mitchell and McElroy, 1946); here adenine riboside is substrate, and the deaminated product is hypoxan- thine riboside, called inosine. Finally the ribose may be hydrolysed off. A third type of reaction occurs with guanosine-5 -phosphate, in which the intact nucleotide is deaminated by *E. coli* B or *S. aureus* without loss of ribose or phosphate to form xanthosine-5 -phosphate (Carter, 1959; Hanka, 1960). With the cell-free enzyme, the equilibrium, in presence of ATP, Mg+% and excess of NH_4 per cent is strongly on the side of animation:

In a variant of this, guanosine-5 -phosphate is *reductively* deaminated to inosine-5-phosphate by *E. coli* in a reaction that is not reversible (Mager and Magasanik, 1960). These reactions allow Aerobacter to convert adenosine to guanosine.

Deamination of the (phosphate-free) riboside and

deoxyriboside of cytosine is carried out by a specific enzyme from *E. coli* (Wang, 1955), producing the uracil derivatives. Direct deamination of the simple purines, as of adenine by the enzyme adenase, is much slower (at least in *E. coli)* than is deamination of the nucleotides. ATP itself is rapidly attacked by *E. coli,* which both dephosphorylates and deaminates it (Lutwak-Mann, 1936).

The pyrimidines, uracil and thymine, are attacked in a different way by a *Mycobacterium species,* being oxidised to barbituric and 5- methylbarbituric acids, respectively. The former is then hydrolysed to urea and malonic acid (Hayaishi, 1955):

> The purines and pyrimidines can also be attacked anaerobically, and here reduction, i.e., hydrogenation, may be the key reaction. From enrichment cultures on orotic acid, (4-carboxy-uracil), an organism called *Zymob. oroticum* was obtained which decomposed orotic acid, anaerobically to CO_2 and NH_3 (Lieberman and Kornberg, 1954- 1955). Evidently hydrogenation is followed by successive splitting of the two peptide bonds thus:
>
> A Corynebacterium found in the same enrichment culture carried out the same reactions except for using TPNH for the first step. The enzyme that converts ureidosuccinic acid, III, to aspartic acid, CO_2, and NH_3 was isolated and found to require Mn^{++} or Fe^{++} The last steps, conversion of aspartic acid to pyruvic acid, CO_2, and NH_3, proceed only in presence of α-ketogluturate and therefore, involve transamination first. Uracil is attacked in a similar way by *Cl. uracilicum* (Campbell, 1957). *Clostridium uracilicum* is a butyric acid fermenter and requires both amino acids and biotin for growth.

The breakdown of the oxygen-containing purines by anaerobic enrichments is of interest in that the products are only NH_3, CO_2, and acetic acid (Barker *et al.*, 1940, 1942, 1944).

Pure cultures of two new clostridia were isolated from the enrichments, using the purine as the sole nutrient, and strictly anaerobic conditions. These organisms, *Cl. acidiurici* (terminal spores with swelling, small rods) and *Cl. cylindrosporum* (larger rods, variable spore position without swelling), are specific for purine breakdown and do not attack glucose, peptones, or amino acids. In keeping with the ability of *Cl. acidiurici* to synthesise acetic acid, it was found also to synthesise riboflavin, folic acid, and biotin and to need no growth factors or amino acids added to the medium (Barker and Peterson, 1944). Such great synthetic power is rare among clostridia.

It should be mentioned in closing that uptake of CO_2 occurs not only in the breakdown, but also in the synthesis, of purines. Growing *E. coli* and Aerobacter incorporate $C^{14}O_2$ into the 6 position of both adenine and guanine, and about 80 per cent of this carbon in fact comes from CO_2 (Koch *et al.*, 1952; Sutton and Werkman, 1953). The adenine and guanine are converted not only to nucleic acids but also to each other.

Fixation of Nitrogen

Although the bulk of animals and plants are unable to utilise in any way the vast stores of nitrogen in the air, there are a number of ways in which atmospheric nitrogen is "fixed" and converted, directly or indirectly, to organic matter. The quantities involved are large: an acre of clover may fix 50 lb of nitrogen per year. In the whole United States, this would amount to a million tons of nitrogen per year. Nitrogen is also being fixed, in increasing amounts, artificially, although the methods require rather drastic conditions. Three such processes are in commercial use: (1) the formation of NO in the electric arc at 3200° C (Birkeland-Eyde process), (2) the production of ammonia by heating nitrogen and hydrogen at about 500° C under pressures of 200 to 300 atmospheres in the presence of metallic catalysts (Haber process), and (3) the formation of calcium cyanamide (CaNCN) by passing nitrogen over red-hot calcium carbide in the electric furnace. The contrast with the smooth quietness of biological fixation, at one atmosphere and 20° C, is very striking.

Fixation by Leguminous Plants

Evidence for Fixation: The first clear evidence that legumes actually absorb nitrogen from the air is due to J. B. Boussingault, the French chemist, who in 1838, on his farm in Alsace, determined the total nitrogen in plants and in their seeds. In pot experiments, to which no nitrogen fertilizer was added, he obtained striking results, from which the following are selected. The legumes thus showed an increase while the non-legumes showed no significant change.

Careful field experiments, in which the quantity and composition of all fertilizers were determined, provided confirmation, and he concluded that "nitrogen may enter the living frame of the plants directly."

However, later on, Boussingault began to be uncertain whether the increase was due to fixation of nitrogen or uptake of ammonia from the air. Liebig, in his celebrated book, *Die Chemie in ihrer Anwendung auf Agricultur und Physiologic* (1840), claimed that Boussingault's nitrogen determinations in his field experiments were unreliable. In later, more rigid experiments, in which the plants were grown in calcined sand or pumice and more precautions taken against contamination, Boussingault (1851-1853) did not obtain any increase in nitrogen with beans or lupines for a reason which will be obvious below. Similarly, Lawes and Gilbert, in field experiments in England, satisfied themselves at first that legumes increased the nitrogen of crops. In a 10- year period, 10 successive crops of wheat totalled 262 kg of organic nitrogen per hectare, while five crops of wheat interspersed with five crops of beans totalled 827 kg of organic nitrogen. Yet later on, Lawes, Gilbert, and Pugh, growing legume plants very carefully on a calcined clay, found no increase in nitrogen. Thus, in each case the increase in nitrogen disappeared when the experimental precautions were increased. On the other hand, another French chemist, Ville, claimed to obtain nitrogen fixation with all plants, legumes and non-legumes alike. This introduced confusion and led to much argument. Approaching the problem in another way, later experiments in Germany showed that fertilizing oats led to increased growth only when the fertilizer contained nitrogen, while peas gave a good response to potassium and phosphate alone (Wagner, 1891). Clarity was reintroduced, though only much later, by Atwater in America, who in 1885 fully confirmed the original findings. The third and fourth lines of data indicate that fixation is repressed by giving large amounts of fixed nitrogen.

In reporting his conclusion that leguminous plants could fix free nitrogen of the air, Atwater admitted that this was "contrary to the general belief and the results of the best investigators on the subject." But it was not long before the complete explanation appeared.

Role of Nodules

At about the middle of the century, considerable attention was directed to the peculiar outgrowths, or nodules, which commonly occur on the roots of legumes. Lachmann in 1858 and Woronin in 1866 observed that the nodule tissue is filled with small rod-shaped bodies, sometimes motile, resembling bacteria. The fact that these organisms, called by Brunchorst in 1885 "bacteroids," are the cause of the nodules was not recognised at that time. A confusing factor was the observation that early in nodule development, the organisms are united in a kind of thread, running through the root hair which is first infected, and then from cell to cell. This thread was thought by some to be a fungus hypha. The exact nature of the organism was only settled by isolation in pure culture (Beijerinck, 1888; Prazmowski, 1890). Beijerinck used for growth medium an extract of leaves of the leguminous plant, together with asparagine, sucrose, and 7 per cent gelatin.

In growing cultures, the nodule organisms are rods, from about 4 Legumes thus differ from most other plants in being able to increase in nitrogen content, and in possessing nodules on their roots. The connection between these two facts was brought to light by Hell- riegel and Wilfarth (1886-1888) in a series of researches of unrivaled clarity. Sand cultures were used, and the plants watered with nitrogen- free distilled water. In brief the findings were as follows:

1. Grass plants grown on sterile, nitrogen-free sand yellowed and made little growth.
2. Pea plants, on sterile sand, were mostly no better. The yellowing and stunting were typical of nitrogen deficiency. No nodules formed. (A few plants, however, became green and flourished.)
3. But if the sterile sand were inoculated with an extract of fresh soil, growth of the peas became good, and the plants were green; on pulling them up, nodules were *invariably* observed.
4. When after inoculation with soil extract the substrate was again sterilised, these effects were not obtained.
5. When young pea plants that had formed nodules were transplanted to sterile, nitrogen-free sand, they continued to make good growth.

It is evident that greening of the plants, i.e., curing of the nitrogen deficiency in the absence of applied nitrogen, is dependent on formation of nodules, and this in turn is dependent on infection from the soil. Correspondingly, the failure of legumes to gain nitrogen in Boussin- gault's later experiments was due to a too careful calcining of the soil and therefore, killing of the nodule-forming organisms.

Gilbert was chairman of the meeting in Berlin at which Hellriegel reported these results, and it was not long before they were widely confirmed by himself and others. Fixation of nitrogen is thus associated with nodule formation. Yet, curiously enough, the nodule bacteria themselves do not fix nitrogen, and fixation must result in some way from the *symbiosis* between plant and bacteria. This is, in fact, the present status of the problem.

Nodules are present on the roots of many non-leguminous plants too. These include *Alnus* (alder), *Coriaria japonica*, *Elaeagnus* and other members of the *Elaeagnaceae*, many of the *Zygophyllaceae*, *Ceanothus* (New Jersey tea), *Myrica*, *Podocarpus*, *Casuarina*, and some of the Cycads. In some of these the causative organisms resemble the bacteria of legume nodules, in others, e.g., *Ceanothus*, it is a streptomyces (Furman, 1958), and in still others perhaps a fungus related to *Plasmodiophora brassicae*, the cause of "clubroot" in cabbage.

The organism of *Alnus* is difficult to cultivate in vitro because it has an absolute requirement for a lipid growth factor present in the *Alnus* roots (Quispel, 1960). It is probably an actinomycete, having "hyphae with a diameter of less than 1 /A." In *Tribulus*, one of the *Zygophyllaceae*, the nodules are not bacterial at all (Allen and Allen, 1950a). Nevertheless, some of these nodules definitely fix nitrogen.

Isolated *Alnus* nodules treated with isotopic N_2^{15} as described on page 381 show fixation which increases for at least two hours (Virtanen *et al.*, 1954). Similarly N_2^{15} supplied to freshly detached nodules of *Myrica Gale* was converted within two hours to amino acids, mostly glutamine (Leaf *et al.*, 1958, 1959). Earlier work on *Coriaria* (Kataoka, 1930) makes fixation probable there also. The ability of *Ceanothus* to flourish on N-deficient soils is often thought significant, but recent attempts to demonstrate its N-fixing ability have been inconclusive (Hellmers and Kelleher, 1959).

Genus *Rhizobium*

The nodule bacteria of different legumes are evidently closely related, and it is reasonable to combine them in a single group or genus. The name *Rhizobium*, proposed by Frank (1890), has been generally adopted. Whether they are a genus with many species, or a species with many strains, is hard to decide and is for all practical purposes meaningless. In their ability to act as antigens on injection into animals, giving rise to antibodies that coagulate members of the group, they show clear interrelations (Kleczkowski and Thornton, 1944).

On the other hand, there is marked specificity within the group in regard to the host plant. Recognition of this fact goes back to Hellriegel's finding that soil in which beans have grown will not support lupines; soil in which lupines have grown will, similarly, not support beans, but it will support serradella *(Ornithopus)*. Cross-inoculation experiments in the field have indicated that there are seven or eight subgroups, each of which is moderately specific.

Numerous other legume species fall outside these groups, and their Rhizobia have sometimes been assigned to additional subgroups. When it is remembered that there are some 10,000 species in the *Leguminosae,* of which barely 10 per cent have been examined (Allen and Allen, 1950b), it will be evident that such a grouping must be excessively tentative. On the whole the first four groups, the "fast- growers," are very similar, while the next three ("slow-growers") are also closely related among themselves.

The cross-inoculation experiments are not always reliable, and when the bacteria have been cultured on nutrient medium, their specificity seems to be reduced. Using cultures in sterile soil, J. K. Wilson (1939) found much less specificity than generally reported in the field; of 37 strains from nodules of *Amorpha fruticosa.* Within a given species, sensitivity is genetically controlled. Thus, one line of red clover *(Trifolium pratense)* has a recessive gene *i* which, when homozygous, causes strain A of *Rhizobium trifolii* to make "ineffective" nodules, i.e., nodules causing little or no nitrogen fixation, while strain CiF and others form effective nodules. In another line derived from this, a recessive factor *m* modifies the first to make strain A "effective." In a third, unrelated line, a different recessive gene *i.e.* makes all strains except CiF ineffective (Nutman, 1959). The sites of action of these factors are different.

The ineffective nodules are usually small and white, the others being larger, pink, and richer in bacteroidal tissue. Crimson clover nodulated by Rhizobium from white clover formed small white nodules and contained 0.4 to 0.5 mg nitrogen per plant, while when nodulated with its own strains, it had 5.3 to 7.3 mg nitrogen per plant at the same age (Burton and Allen, 1950).

In other groups, especially III and V, flagellation depends on the stage of growth, the smallest coccoid cells being monotrichous, or occasionally with two polar flagella (Bisset and Hale, 1952). An organism called *B. radiobacter*, named from its tendency to form radiating, star-shaped groups, resembles Rhizobium in many ways and can be found on the roots of legumes, but does not form nodules or fix nitrogen.

It is the commonest contaminant of Rhizobium isolates, but can be distinguished by its ability to grow in yeast extract-mannitol broth at *pH.* 11 and to produce acid from dextrin, inulin, and amygdalin in yeast extract media (Hofer, 1941). The organism of crown-gall disease of plants, *Ps. tumefaciens,* also forms star-shaped groups, and the galls formed by it in some plants superficially resemble nodules. Both these organisms have been included by Bergey in the oddly named genus Agrobacterium which is placed next to Rhizobium in the family *Rhizobiaceae.*The key description of this is as follows: "Heterotrophic rods which may not require organic, nitrogen for growth. Usually motile with 1 to 6 or more flagella. Usually form nodules or tubercles on roots of plants, or show violet chromogenesis." The description is, of course, absurd. As to organic nitrogen, Rhizobia require it, Agrobacterium does not, so that the word '"may" means only that this criterion does not apply. The alter native between nodule formation and violet chromogenesis does not suggest any particular unity in the group, apart from the fact that *B. radioltacter* does neither all, the Rhizobia are probably more nearly related to the Pseudomonads than to any other group.

The "bacteroid" form is thought by some to be connected

with nitrogen fixation, and in some types of ineffective nodules the rhizobia are rod-shaped (Virtanen, 1947; Nutman, 1959). In culture, bacteroid forms can be induced by adding alkaloids, metallic chlorides, or malic acid (Fred, Baldwin, and McCoy, 1932). Sometimes bacteroids are seen in young cultures, particularly when growth is slow. The bacteroid shape might point to the Actinomycetes, and it is suggestive that the symbionts of *Alnus* and other non-legumes are often placed in this group, but the motility and gram- negativity of Rhizobium are against such a relationship.

The barred staining of the bacteroids has been ascribed variously to fat bodies (Lewis, 1941), to cross walls with cytoplasm aggregated on them (Bisset and Hale, 1952), or to their being a kind of sporangium, the darkly staining granules being released as small cocci or "swarmers," which afterwards grow again into rods.

This "life-cycle" theory might seem supported by the variations in morphology which Rhizobia show, because small coccus-like forms are often seen together with rods of varying length, and perhaps the large barred rods divide into very small cells. Individual bacteroids, when in single-cell isolations, are quite incapable of growth (Almon, 1933), and since the irregularity of staining is in part due to degeneration of the nuclear material (Bergersen, 1955), it seems likely that they are truly decadent forms. Some viable forms must, of course, remain in vivo, or nodules would not yield cultures.

Physiology of Nodule Formation

The infection process was first described by Marshall Ward in 1887 and by Frank in 1890. The bacteria usually accumulate at the tip of the root hair, which they enter, causing curling or other deformation. The bacteria alone do not produce cellulase or pectinase to attack the root-hair wall (Clarke and Tracey, 1956), but in combination with a clover seedling they do (Ljunggren and FShraeus, 1961); it appears that the Rhizobia secrete something that stimulates the root hair to produce a pectinase.

(The tip of the root hair contains little or no cellulose.) Within the living root hair some separate motile bacteria can at first be seen (Burgin-Wolff, 1959), but the bacteria soon agglomerate into a thread which penetrates from cell to cell in the root cortex. This infection thread is a peculiar structure; it has a definite membrane of cellulose and also contains pectin; evidently, therefore, it is contributed by the host plant and riot by the bacteria that are inside it. In its origin it can be considered as an invagination of the original root-hair wall (Nutman, 1956, 1961), and ultimately it comes to surround each subgroup of rhizobia in the cells. At each cell wall, the thread probably passes through the primary pit areas, which have a loose, open structure, doubtless made more so by the pectinase secreted; the rods remain lengthwise in the thread.

Development of the nodule tissue depends on presence of the few tetraploid cells that each root normally contains. When these are entered, the nucleus is stimulated to divide, so that the amount of bacteroidal tissue increases.

Since the diploid cells do not respond, the infected core of the finished nodule is largely tetraploid and its cortex diploid (Wipf and Cooper, 1940; Fujita and Mitsuishi, 1953). The infection thread seems to be attracted to the nucleus in each cell, and the same is true in the root hair; if infection occurs on one branch of a forked root hair and the nucleus is in the other branch, the thread will turn outward towards the nucleus rather than inward to the cortex (FShraeus, 1957).

Within limits, the number of nodules formed is a function of the number of rhizobia, a minimum of 10^6 to 10^9 per cubic centimetre, of rhizosphere medium being needed for infection (Purchase and Nutman, 1957). But beyond that, several factors control nodulation; just which ones are crucial is still not known. In the first place, only a few per cent of infected root hairs give rise to nodules. This is in part, of course, due to the strain specificity; in Burgin-Wolff's observations on living root hairs (1959), the root hairs of peas could be infected by rhizobia from clover, and even sunflowers and corn had some rhizobia in their root hairs, but of course no nodules resulted. But even with pure strains of the host-specific type, one clover species may form a nodule from each infection, while another forms only one from 100 infections (Nutman, 1961). Obviously many infection threads abort.

In the second place, there is an interrelationship between nodules and the primordia of lateral roots. In clover, young uninoculated plants bear more laterals than inoculated ones, the total number of laterals plus nodules remaining constant for a time. Decapitation of the root tip increases nodule formation, perhaps by increasing the number of lateral root initials. Nodules themselves inhibit the formation of other nodules, and the inhibition is exerted by the uppermost part which contains the meristem; when nodules (or their meristems) are cut off, the rate of infection increases until the original number is about reinstated (Nutman, 1952).

All these findings indicate that the formation of nodules and of lateral root initials is subject to a predetermined *pattern* in the host's root system which is mediated by the secretion of inhibiting substances. It is known that lateral roots are subject to such a pattern of control (Torrey, 1952, 1956), and thus, as Nutman (1961) concludes, "the nodule functions essentially as one meristem among many others active in determining the pattern of growth." The diffusible substances responsible for this interrelationship may be internal or external. Seedlings grown in a medium in which another seedling has previously grown and been removed become nodulated earlier than controls. On the other hand, if many seedlings are together in a limited volume, the number of nodules is decreased (Nutman, 1956). Addition of fine absorbent charcoal increases the number (Turner, 1955), i.e., secretions can be both promotive and inhibitive.

Internally, factors come from the cotyledons which control the formation of nodules and of lateral roots (Thornton, 1929; Torrey, 1956, respectively), and since inositol fed through the root base increases nodulation (Raggio *et al.*, 1959), it may itself be one of these factors. If the number of tetraploid cells is small, this may limit nodulation too. Treatment with colchicine (which doubles the chromosome number) greatly increases the number of nodules on diploid clover and serradella, but has a smaller effect on tetraploid strains (Trolldenier. 1959). This might explain why effective and ineffective strains compete in nodule formation, as they have long been known to do, because clover plants inoculated first with an ineffective strain derive little or no benefit from subsequent inoculation with an effective one, and vice versa (Burton and Allen, 1950).

The various stages of nodulation offer numerous points at which the process can be interfered with. For example, Nutman's factor *i*, when homozygous, prevents the conversion of the rhizobia into bacteroids; the bacteria multiply but remain normal, and the typical degeneration of the nucleus does not occur. The factor on the other hand, prevents multiplication, so that the bacteria remain in a little clump at the point of discharge from the infection thread.

Their secretion (indoleacetic acid) stimulates cell division in the normal way, however, so that a tumour, with little or no bacterial tissue, develops (Bergerson and Nutman, 1957). In another clover, *T. subterraneum,* one strain of Rhizobium cannot escape from the infection thread but multiplies within; the resulting nodules are no more than tiny bumps.

Thus the control of nodulation seems to be more in the hands of the host plant than of the parasite. When we consider also that the plant contributes a wall to the infection thread as well as (probably) an enzyme to attack its own walls, and that it has to nourish the Rhizobia throughout and eventually hydrolyse them, it is clear that the symbiosis is elaborately developed. Some authors claim that the relationship is not limited to the nodules.

Frank (1890) and others later have claimed that Rhizobia penetrate to the upper parts of the plant, and although this has been heartily denied (Allen and Allen, 1950b), it does not seem any more improbable than the penetration of distant organs by crown-gall bacteria. Tonzig and Bracci (1951) conclude that "legumes normally harbour abundant Rhizobia, not only in the nodules and root-system but in ... the stem, leaf, different parts of the seed and even, very probably, parts of the flower." In any event, it is only in the nodules that the plant responds with a growth reaction.

Growth of the nodule, once it is initiated, is associated with, and perhaps caused by, an excess of indoleacetic acid. It was the finding that nodules, both of legumes and of *Myrica*, are rich in an auxin that led to the proposal that they represent lateral roots whose outgrowth is inhibited in length and increased in diameter by auxin (Thimann, 1936, 1939). These are known effects of auxins on roots. The fact that the nodule is cortical while the lateral root is initiated on the edge of the central stele is, however, an important difference (Bond, 1948), and it is more probable that both represent points of *potential* lateral outgrowth. Not only are nodules far richer in auxin than ordinary root tissue, but their auxin is all in the free, or active, form, while in roots, it is only slowly liberated by enzymatic action (Thimann *et al.* 1942). The curling of root hairs so characteristic of the initial stages is also caused by auxin. Rhizobia produce quantities of indoleacetic acid in culture media also (Chen, 1938; Thimann, 1936, 1939).

The auxin in Phaseolus nodules increases as they enlarge to reach a maximum at the end of flowering, then decreases as the pods grow and the nodules correspondingly begin to degenerate. Thus, the auxin is probably the major cause of the nodule's growth. There is also differentiation, since vascular bundles soon develop near the outside of the nodule and establish contact with the bundles in the root. This ensures a good supply of carbohydrates to the nodule and no doubt carries away amino acids to the shoot (Pfennig, 1956). The vascular bundle of the lateral root, in contrast, develops as a branch from that of the main root (Bond, 1948). The core of bacterial tissue normally becomes pink due to haemoglobin, and when the nodules degenerate and their auxin content falls, this turns greenish. The bacteroids then disappear, the nitrogen remaining in the nodule is absorbed into the plant like that in leaves before their fall in the autumn, and finally the nodule is cut off by a cork layer and withers away. In some perennial lupines, however, bacteroidal tissue becomes enclosed within the root and produces a new nodule the next season. What happens in the tropical leguminous trees is not known.

Electron-microscopic study of developing nodules has brought out another suggestive anatomical point (Bergerson and Briggs, 1958). When the bacteria are released from the infection thread, each is surrounded by a fragment of the thread membrane, and as each one divides, the membrane enlarges, so that the resulting group of daughter cells is enclosed in a thin membrane.

All the rhizobia in the nodule are thus in little membrane-enclosed units, usually four to six in each one. This probably means that the microenvironment of the rhizobia is still more anaerobic than that of the host cells.

Mechanism of the Symbiotic Nitrogen Fixation

Since neither the Rhizobia alone nor the legume alone will fix nitrogen, the problem of the fixation mechanism is very difficult. Isolated nodules will fix nitrogen, but only in small amounts and for a few hours.

A system intermediate between isolated nodules and intact plants is offered by nodules in culture on isolated root systems, fed with nutrient medium; these have a normal pink colour and fix nitrogen in approximately normal amounts (Raggio *et al.*, 1957, 1959). The problem is in part one of nutrition, because fixation in whole plants is greatly reduced by darkening the plants for a few hours, although it continues slowly in the dark (Lindstrom *et al.*, 1952). The Wisconsin workers at one time assembled much evidence to indicate that nitrogen fixation and the extent to which nodules are formed are dependent on the carbohydrate: nitrogen ratio in the plants. Thus, increasing the CO_2 content of the air or adding sugar to the soil, both of which increase the carbohydrate in the plants, increases nitrogen fixation and nodule number. Conversely, manuring with high nitrogen inhibits both processes. The conception unfortunately does not shed much light on mechanisms.

Nitrogen fixation is inhibited by hydrogen gas. When combined nitrogen is supplied, hydrogen does not affect the growth of the plants, but when they are growing without added nitrogen and dependent on fixation, hydrogen strongly inhibits growth and fixation. If we assume that N_2 is taken up by a special enzyme, *nitrogenase,* then its K_m, i.e., the N_2 concentration giving 50 per cent saturation, is for soybean nodule slices about 0.025 atmosphere of N_2 (Burris *et al.* 1955). This value agrees with the K_m found for Azotobacter and Nostoc. Hydrogen, on this basis, would combine at the same enzyme and thus displace N_2; its affinity for the enzyme is about one-fifth that of N_2 (Wilson and Burris, 1953).

Carbon monoxide also inhibits fixation, 0.0001 atmosphere of CO having a detectable effect in red clover. However, unlike the inhibition of cytochrome oxidase, the effect of CO here depends only on the CO pressure and not on the $CO:O_2$ ratio. It might be deduced that nitro- genase is not a cytochrome but a related enzyme of the hemin-protein type-combined nitrogen, as we have seen, inhibits fixation too, but in a different way. Since the infected root hairs become curled or deformed, the early stages of infection can be followed by determining the numbers of curled root hairs; this number is decreased by adding *nitrate.*

For example, with alfalfa (Medicago) in sterile culture, 13 per cent of the root hairs were deformed; when Rhizobia were added, 64 per cent were deformed; if 0.15 per cent $NaNO_3$ was added as well, the number fell to 16 per cent (Thornton, 1936).

If nitrate was supplied to isolated roots through their bases, nodulation was not decreased; only if it reaches the fine roots *externally* is nitrate inhibitory (Raggio, Raggio, and Torrey, 1957). Thus, the effects of nitrate and other oxidisable forms of nitrogen are probably exerted via this first stage rather than on the No-fixing process.

In 1933, Virtanen and co-workers in Finland discovered that legumes grown in nitrogen-free sand excrete organic nitrogen compounds into the sand (Virtanen *et al.* 1933). The excretion is large enough in amount to allow the growth of some grasses (or even spruce-tree seedlings) in the sand. It appears to be a true excretion rather than merely a sloughing off of decayed nodules. Most of the nitrogen appearing in the sand is in the form of aspartic acid; with smaller amounts of its decarboxylation product, 0-alanine, and of glutamic acid (Virtanen and Laine, 1939; Virtanen, 1947).

This would imply that the fixed nitrogen had become combined with oxalacetic or ketoglutaric acid. It had earlier been shown that clover without nodules grows well with aspartic acid as a sole nitrogen source. The findings were disputed at first, but were later repeated in the United States under growth conditions similar to those prevailing at Virtanen's laboratory in Finland. There was some evidence that the first fixation product was hydroxylamine, but this was subsequently shown to be improbable both by the Wisconsin group and by Saris and Virtanen (1957).

However, the positions of aspartic and glutamic acids as first products have been fully supported. Virtanen's original claim that isolated nodules will fix nitrogen in the presence of oxalacetic acid has not been borne out, because healthy soybean nodules separated rapidly from the parent plant fix nitrogen (supplied as N_2^{15}) only for a few hours, and the period is not prolonged by supplying oxalacetate or any other organic acid (Aprison *et al.*, 1954).

Nitrogen fixation by excised Alnus nodules is also not much increased by organic acids (Virtanen *et al.*, 1954). Either the keto acid must be produced endogenously to be effective, or else the system is very fragile. Glutamic acid appears as a primary product of fixation by the free-living Axotobacter and

Clostridium. Also, in nodules of the non-leguminous tree *Myrica Gale* (bog myrtle), after incubation with N_2 most of the isotope was found in the amide N of glutamine and asparagine (Leaf *et al.*, 1959). In *Alnus* nodules the N_2^{15} goes into glutamic acid and the guanidine N of citrulline (Leaf *et al.*, 1959).

Although there is thus uncertainty about the exact pathway, and perhaps some interspecific differences, this conception does indicate why fixation is so dependent on the plant's nutrition and why isolated nodules fix only transiently. For if a supply of keto-acid precursors must be maintained, the nodule must be in constant touch with the photosynthetic system. The vascular bundles that supply the nodules are thus very important, both for supply and for product removal.

Since nodulated and unnodulated roots of peas have the same amino acid composition (Pfennig, 1956), the fixed nitrogen is probably continuously given off to the rest of the plant. In boron deficiency, vascularisation of the nodules is greatly reduced, and correspondingly growth and fixation are seriously decreased. A remarkable fact about nitrogen fixation in nodules is its dependence on traces of molybdenum and cobalt. Molybdenum has long been known to control fixation by free-living bacteria, but recently this was extended to nodules (Bond and Hewitt, 1961), thus confirming a strong suspicion of some 30 years' standing.

A similar requirement for cobalt is shown by several legumes and by Alnus (Ahmed and Evans, 1961; Bond and Hewitt, 1962). When these plants are grown in purified nitrogen-free sand culture and are inoculated, they become yellow and appear N-deficient unless 1 part in 10^8 of cobalt is added; the N content of the plants is then tripled. The same appears to hold for Casuarina.

Some suggestive observations on pigments may bear on the first stage of fixation. It is well known that healthy nodules are usually pink, and the term "beef-steak" has even been used in describing them. As it happens, the term is unexpectedly appropriate because the pigment is actually haemoglobin. It combines with O_2 and with CO to form an oxyhemoglobin and a carboxyhaemoglobin, just as does the haemoglobin of blood cells. It readily oxidises to a brown form in which the iron is trivalent methaemoglobin. In older nodules, or in plants kept in the dark for a few days, it turns greenish. Blood haemoglobin undergoes a similar change when it is metabolised, forming a green bile pigment. Virtanen (1947) points out that the change to green is invariably associated with loss of nitrogen-fixing power in the nodule.

Nodules caused by ineffective Rhizobium strains are almost colourless, as shown by Thornton's (1945) observations with several hundred strains. Their content of total haematin is about one-tenth that of effective nodules (Smith, 1949), and the *brei* from ineffective nodules does not form any porphyrins in vitro, under conditions where the *brei* from effective nodules does (Falk *et* al., 1959). Rhizobia in culture can form porphyrins, although they never form haemoglobin (Falk *et al.*, 1959). Neither is haemoglobin ever found in higher plants, except in nodules. It is suggestive that the nitrogen-fixing capacity of nodules varies with the effective volume of haemoglobin-containing tissue (Ber- gersen, 1961).

These facts suggest that the haemoglobin might combine with nitrogen, as the first step in the process. Indeed, observation of the spectrum of soybean-nodule extract in an atmosphere of helium shows the haemoglobin mainly in the ferrous form, and on admission of N_2 it changes to the ferric form (Hamilton *et al.*, 1957). Hydrogen causes the reverse change. This certainly indicates some reaction with nitrogen. However, addition of nodule haemoglobin to Rhizobia in culture does not cause any nitrogen fixation (Tove and Wilson, 1948); added oxaloacetate does not help. The haemoglobin of blood does not combine with nitrogen either, although the two haemoglobins are not necessarily identical.

Indeed, Virtanen's preparation (called "leghaemoglobin") appears to have a molecular weight (17,000) about one-quarter that of the blood pigment. The haemoglobin of nodules definitely has no function in respiration; effective and ineffective nodules have the same rates of O_2 uptake, and when all the haemoglobin is combined with CO (through exposure to a mixture of CO and air), the 0_2 uptake is not reduced (Smith, 1949). It is, however, suggestive that blood haemoglobin combines with hydroxylamine, liberating free N_2 (Colter and Quastel, 1950), but there is as yet no evidence that the reaction is reversible. Thus, the physiological function of the haemoglobin remains obscure, and with it, the whole mechanism of nitrogen fixation in nodules.

Fixation by Free-living Organisms

Clostridium: In formulating his well-known "Law of the Minimum," according to which the yield of a crop is determined by that element which is present in the soil in minimum (i.e., limiting) amount, Liebig, 100 years ago, recognised that any one of many elements might, under different conditions, be limiting the crop. He considered nitrogen one of these and urged the use of ammonium salts as manures. The need for constantly replacing nitrogen was established experimentally by the classic researches of Lawes and Gilbert in 1843-1855, who found that all plants, with the exception of legumes, need nitrogen added to the soil.

In 1847, Liebig changed his views, concluding that meadow land supplied with alkali and phosphate could actually gain in nitrogen. The basis for this view is not clear, since Boussingault's careful experiments showed that plants in general could not gain nitrogen from the atmosphere. However, certain experiments of Berthelot (1885) cast a different light on the matter. Samples of sterilised and unsterilised soil were set aside in the laboratory or outdoors for some months. The unsterilised samples were found to have increased, slightly but definitely, in total nitrogen. The increase was almost entirely in the organically combined nitrogen.

A few years later, Winogradsky made his well-known

study of nitrogen fixation (1895). A culture medium free from nitrogen but containing sugar, phosphate, sulphate, and $CaCO_3$ was inoculated with soil. Vigorous fermentation took place, with the smell of butyric acid. After a time the solution had gained in nitrogen.

In the course of numerous transfers to fresh media, it was soon found that sporeforming rods constituted the major part of the flora. Winogradsky therefore, pasteurised the solution to remove non-spore- formers. Furthermore, the sporeformers grew mainly at the bottom of the culture, suggesting that they were anaerobes. He therefore, used tall cylinders through which nitrogen was bubbled. In this way, he finally isolated a sporeforming anaerobe, which was named *Clostridium pasteurianum.*

TABLE: Berthelot's Experiments on the Increase in Soil Nitrogen (Data from Berthelot, 1885)

Total N In Mg/Kg Dry Weight				
Treatment	*Determined In:*			*Average Gain (mg/kg) per 3 Months*
	April	*July*	*October*	
Pots indoors	98.9	116.7	120.9	11 0
Pots in the field	26.9	45.8	63.0	18.1
Pots sterilised, open to air through cotton plugs: 1		116.7	109.2	0
2		150.2	134.8	0
3		35.4	32.2	0

Isolation of an actual nitrogen-fixing organism settled the matter, insofar as gains of nitrogen by soil are concerned. It also caused a stir among agriculturists, who saw the possibility of inoculating soil with this wonderful organism and thus ending forever the need for supplying nitrogenous fertilizer. Within a year or two, Caron placed on the market a culture to increase fixed nitrogen, named "alinite." This was useless for two reasons. First, it was not *Cl. pasteurianum* at all, but an aerobe, *Bac. cereus*, common in soil and, like other bacilli, quite without nitrogen-fixing power. Second, as we know now, the flora of soil is usually determined by the prevailing conditions and not by inoculation.

If soil contains attackable carbohydrate and the right salts, the clostridia will develop in any event as soon as conditions become anaerobic enough. Bredemann (1908) found *Cl. pasteurianum* in 137 out of 152 soils examined. Those from which it was absent comprised mainly the peats and peat-sand soils with more acid than 5. In Winogradsky's work with soil it was enough to add glucose and make the soil waterlogged to develop enormous numbers of *Cl. pasteurianum* immediately.

Experiments like those of Berthelot have recently been repeated in Nigeria (Moore and Abuelu, 1959); the gain was 28 mg N per kilogram of soil in five months. Addition of Mo and phosphate to the pots of soil more than tripled the gain, which reached 110 mg per kilogram.

The power of fixing nitrogen is widespread among the Clostridia, with the exception of those that are animal pathogens. As with Rhizobium, the amount of nitrogen fixed depends on the amount of carbohydrate supplied.

Winogradsky (1902) found *Cl. pasteurianum* to fix 2 mg N per gram of dextrose fermented, but later workers with shaken cultures have obtained up to 7 mg N per gram of dextrose, with *Cl. saccharobutyricum* and other species. Even so, a gram of sugar gives rise to less fixation with Clostridia than it does with Azotobacter, but since the sugar has to be fermented rather than oxidised, it yields much less energy per gram.

In nodules and aerobic organisms, the fixation is inhibited by ammonia, but *Cl. pasteurianum* can use both N_2 and NH_3 simultaneously, because when the cells are exposed to N_2^{15}, it is converted to $N^{16}H_3$, which appears in the medium. In other respects, the action of inhibitors is similar to that in aerobes. Hydrogen inhibits, the inhibition being proportional to the ratio H_2 N_2, i.e., it competes with nitrogen (Hiai *et al.*, 1957);

both CO and N_2O inhibit also (Hino, 1955). It is suggestive that CO inhibits ammonia assimilation too, although less powerfully than it does nitrogen fixation.

Myxophyceae (Blue-Green Algae)

Nitrogen can also be fixed under aerobic conditions. This was brought out by Beijerinck (1901, 1902) in a study of the flora of tap water containing 0.02 per cent KH_2PO_4. Organisms growing in the presence of only traces of nitrogen he called *oligonitrophilic* (Greek *oligos,* small or few). This medium, inoculated with soil and kept in the light, slowly develops blue-green algae, principally the filamentous forms, Nostoc and Anabaena. Only if ammonium salts were added did grass-green algae *(Chlorophyceae)* appear.

Beijerinck, therefore, concluded that the blue-greens could fix nitrogen, and much later this was confirmed by nitrogen determinations (Drewes, 1928; Fogg, 1942); all filamentous types that form heterocysts can fix N (Wilson and Burris, 1953). The amount of nitrogen fixed was originally considered to be very small, Drewes finding only 12 mg N per litre of medium in 50 days; however, it is greatly increased by vigorous shaking to bring the cells in contact with the N_2. *Anabaena cylindrica* in shaken cultures is reported to fix up to 500 mg N per litre (Allen and Arnon, 1955).

The large range in the recorded amounts of fixation may be due to contamination by bacteria. Thus *Nostoc calcicola* in pure culture fixed only 10 mg N per litre of medium in 90 days (cf. Drewes' value), but in mixed culture with *Bac. megaterium* 54 mg N were fixed (Bjalfve, 1962).

The fixation by blue-greens is inhibited by H_2 and by CO to about the same extent as that by Azotobacter and by clover nodules. Since blue-green algae are of very wide distribution in lakes and streams, the fixation may be quantitatively important there; they also grow vigorously in flooded rice fields. A number of lichens have Nostoc as the algal partner, and two of these, as well as the liverwort *Blasia pusilla* which contains Nostoc in the tissue, have been shown to fix small amounts of nitrogen (Bond and Scott, 1955). A few marine blue-greens have been found to fix nitrogen also (Watanabe *et al.*, 1951) which again indicates the probable large scale of the process in nature.

Purple Bacteria

In studying the photosynthetic reactions of Rhodospirillum, a purple bacterium, it was found that from some organic acids hydrogen is given off in the light (Gest *et al.*, 1950). The reaction is inhibited by nitrogen gas, as well as by ammonium. Because of the special relations between hydrogen gas and nitrogen fixation, it was thought that the reciprocal relationship might apply here, and that the Rhodospirillum might be able to fix nitrogen. Accordingly, the culture was tested, using N^{15} gas, and found to fix nitrogen vigorously in the light.

Nitrogen gas could even be used instead of amino acid as source of nitrogen for growth, provided a trace of biotin were added; after several weeks, 56 mg N_2 was fixed per litre of medium (Gest *et al.*, 1956). This is about five times as much fixation as by Nostoc.

Under similar conditions, but without biotin, both a Chromatium (of the *Thiorhodaceae*, or sulphur purples) and a Chlorobium have been found to fix up to 100 mg N_2 per litre after two to three weeks (Lindstrom *et al.*, 1950).

Nitrogen is also fixed by 19 strains of Rhodopseudomonas, representing all four species, and by Rhodomicrobium (Lindstrom *et al.*, 1951). Hence, the ability to fix nitrogen, and in considerable quantities, is characteristic of the *Rhodobacteriinae* as a whole. Fixation by Rhodospirillum is greatly promoted by sulphide, and, as with other organisms, is inhibited by H_2 (Gest *et al.*, 1956).

Ordinary Bacteria

In the last few years bacteriologists have come to realise the surprising fact that many "ordinary" bacteria have the ability to fix a little nitrogen. These weak nitrogen-fixers include: *Aerobacter aerogenes Methanobacterium omelianskii "Ps. azotogenesis,"* a yellow Pseudomonas from soil, and six other Pseudomonads *Achromobacter*, eight strains, and *Nocardia calcarea* with the Methanobacterium, which produces CH_4 from alcohols, traces of ammonia actually promoted the fixation by promoting growth, but larger amounts inhibited strongly (Pine and Barker, 1954). The Achromobacter (Jensen, 1958) fixed 3 to 17 mg N per litre of medium in four days, at a rate of 1 mg N per gram of carbohydrate used, which is considerable, though much less than with typical nitrogen fixers.

It could even fix some nitrogen anaerobically. Some of the Pseudomonads fix as much as 50 mg N per litre of medium (Proctor and Wilson, 1958-1959). They respond to small amounts of ammonium by increased growth, like Methanobacterium, but then the growth slows down and after a lag it restarts, this time with N fixation. This suggests that NH_3 represses formation of the nitrogenase enzyme.

In addition to the bacteria, two red yeasts (Rhodotorula), a Sac- charomyces and a black Pullularia, all isolated from soil, fix moderate amounts of nitrogen (Metcalfe and Chayen, 1954; Metcalfe and Brown, 1957; Nemeth, 1959). The Pullularia fixes 4 to 5 mg N per gram of glucose supplied, which is comparable to the Clostridia. Since up to 8000 were found per gram of soil (Metcalfe and Brown, 1957), its contribution may be important. One wonders how many other nitrogen-fixing organisms will be turned up.

Azotobacter

1. *Characteristics:* If carbohydrate be added to a medium containing salts but only traces of nitrogen, many bacteria develop aerobically. Of these, one of the most characteristic is Azotobacter, a nitrogen-fixer of the first importance. Beijerinck and van Delden (1902), who isolated it, could obtain fixation only in mixed cultures, but later workers have shown that pure cultures fix

nitrogen very actively. The amount is large up to 20 mg per gram of sugar (Fischer, 1949; Wilson and Burris, 1953), or more than double that of most clostridia. In addition to Beijerinck's original two species, *A. chroococcum* and *A. agilis,* several others have since been discovered.

The Azotobacters have been described in detail by Jensen (1958), who considers that *Az. indicum,* because of its long lag period, slow growth, and formation of acetic acid and excessive amounts of gummy polysaccharide, should be placed in a new genus, *Beijerinckia.* On a glucose medium it converts 2 to 3 per cent of the glucose to acetic acid, 30 to 70 per cent to gum, and the rest to CO_2 (Quinnell, 1957). Each cell contains two prominent fat globules. A smaller-celled type with numerous fat globules, and clearly polar flagella, has been placed in a third genus, *Derxia* (Jensen *et al.*, 1960; Roy and Sen, 1962). The other species are fairly closely related, although within the Chroococcum type, the initial rate of nitrogen fixation indicates two subgroups (Fischer, 1949), since of 68 strains, about half fixed 2 to 4 mg N, and the others, 8 to 12 mg N (in 2 days at 30°C on 50-ml mannite medium). Very few were intermediate. The upper value, 240 mg per litre, is not as high as has been recorded. The more active group was the more motile and more nearly rod-shaped. Azotobacter cells have been thought to resemble those of blue-green algae. The relatively broad cells, showing characteristic granules, are easily recognised in enrichment cultures. Their dimensions, unusual in bacteria, correspond to those of small blue-greens such as *Chroococcus dispersus* var. mmor, which is about 2 in diameter. The gummy capsule holds the Azotobacter cells together in pairs or small groups. In addition, single cells form curious cysts, with at least three layers of wall. These might be compared to spores, but the cell inside shows no differences from its normal vegetative form (Tchan *et al.*, 1962) and oxidises substrates vigorously (Socolofsky and Wyss, 1961). The cyst wall is thus like a capsule which has hardened. The fact that all Azotobacters are motile with peritrichous flagella is an important difference from the blue-greens. Unfortunately also for the parallel, *Chroococcus* and the family *Chroococcales* are not among the blue-greens known to fix nitrogen (Fogg and Wolfe, 1954).

2. *Metabolism and the Mechanism of Nitrogen Fixation in General:* The Azotobacters are strictly aerobic. Their respiration is fully cyanide-sensitive and is about as rapid per unit of protein as that of any known cell. Theirs is an atypical cytochrome system, and they were used for the direct observation of the bands of the oxidase, cytochrome a_z, which replaces cytochrome *a* in these cells. The bands of *d* and b_l take the place of cytochromes *a* and *b*. In *Az. vinelandii* (we do not know if the same is true in others) cytochrome *c* is replaced by the unusual pair C_4 and C_5, which have not been found to interact with the cytochrome systems of other organisms (Tissieres, 1956). Thus, a peculiarity of the Azotobacters, comparable with their morphology, is their cytochrome system. It is suggestive that the purple bacteria also have unusual cytochromes, and it would be worth while to know more about the cytochromes of Nostoc and Rhizobium.

As far as substrate is concerned, Azotobacters are not highly specific. Organic acid salts, sugars, and sugar alcohols are favourable for growth, and mannitol is often used in enrichment cultures because not many Clostridia can attack it; Winogradsky's observation of the rapid growth of large cocci when mannitol was added to soil doubtless refers to Azotobacter.

With a variety of carbon sources, the nitrogen fixation is roughly proportional to the heat of combustion of the compound (Mockeridge, 1915), which probably means that it is proportional to the amount of combined hydrogen present. The rule does not hold for benzoate (Fischer, 1949). Of great importance for the fixation process is molybdenum, a concentration of 1 in 10^{10} having a marked effect (Bortels, 1930, 1936).

However, *Az. vinelandii* and *Az. agile* can fix nitrogen in a highly purified medium about half as fast without Mo as with it (Bove *et al.*, 1957); only *Az. chroococcum* fixed no nitrogen without added Mo. In many cases, but not all, molybdenum is replaceable by vanadium but by no other metal. The Mo is antagonised by tungsten (as lung- state), giving an apparently competitive inhibition (Takahashi and Nason, 1957), and tungstate inhibits the growth of Azotobacter when N_2 gas is the nitrogen source (Aurich, 1959), although this may be only because it prevents the uptake of Mo (Keeler and Varner, 1957). Vanadium does not reinstate growth in presence of tungstate. Most of the Mo taken up has been found, by using the isotope Mo^{90}, to go into a protein-rich fraction (Keeler *et al.*, 1956; Magee and Burris, 1956). Even in *Aerob. aerogenes* Mo is required for nitrogen fixation, 1 part in 10^8 giving maximum effect; there was no Mo requirement for growth on an ammonium medium (Pengra and Wilson, 1959). Most higher plants require Mo for growth, but legumes fixing nitrogen probably require more than others, and nitrogen fixation in nodules appears dependent on Mo (Bond and Hewitt, 1961). The reduction of nitrate also requires Mo. All in all, it can be deduced that the nitrogenase enzymes in virtually all organisms contain molybdenum. In Rhizobium, cobalt promotes growth, both of the bacteria and of legumes growing symbolically with them; it is converted to vitamin B_{12} co-enzymes in the cell and in the nodule (Ahmed and Evans, 1960-1961; Kliewer and Evans, 1962). However, a direct action of Co on fixation is doubtful.

As mentioned in connection with nodules, an early candidate for the position of first intermediate in nitrogen fixation was hydroxyl- amine. This compound was also considered in relation to Azotobacter, but it is quite toxic; N^{15}-labelled hydroxylamine is not used by the bacteria, and oximes are assimilated only poorly (Burris, 1956). Since the starting material is N_2, an intermediate containing 2 N atoms is obviously attractive, and some interest has centred on hydrazine, N_2H,. Hydrazine, although it is also toxic, is readily taken up by Azotobacter, and ammonia inhibits the uptake. Chemically, hydrazine is known to react with α -ketoglutarate.

Thus, the evidence does not support either NH_2OH or NH_2NH_2 as an intermediate, and ammonium remains the strongest contender for the title. In its favour is the fact that Azotobacter and Clostridium use it very rapidly and convert it to the same products as they do N_2. After three minutes' exposure of Azotobacter to $N^{15}H_4+$, the highest concentration of N^{15} was found in glutamic and aspartic acids, and the same was true after 90 minutes' exposure to N_2^{15} gas. Unlike most other amino acids, these two compounds are assimilated by Azotobacter when supplied exogenously.

Cell-free preparations convert ammonia into the same products, *Az. vinelandii* extracts incorporating $N^{15}H_4^+$ into glutamic acid without delay (Burma and Burris, 1957). Addition of a-ketoglutarate accelerated the NH_4^+ uptake, as might be expected. It was shown by Winogradsky (1930, 1938) that when Azotobacter grows on sodium lactate or benzoate, its oxidation of the organic anion sets free so much alkali that the cells autolyse and ammonia is evolved. The spectacle of enough NH.< coming off a *nitrogen-free* culture to turn red litmus paper blue is certainly dramatic. Pertinent to the role of NH_3 also is the long-known fact that ammonium salts (and urea) inhibit fixation, while amino acids have little effect. Against a *primary* role of NH_3, however, are the data with isolated nodules exposed two hours to N_2 (Aprison *et al.*, 1954), where most of the N^{15} was found in glutamic and aspartic acids, and the concentration of N^{15} (= "specific activity") was *higher* in these than in the NH_4. Azotobacter in exposures of two to five minutes to N_2^{15} shows the same reaction (Allison and Burris, 1957). These data suggest rather that NH_4 is secondarily formed (perhaps very easily) from the primary fixation product.

Most of the recent work has centred on fixation by cell-free systems. Protoplasts from *Az. vinelandii*, made by treating with lysozyme and versene, and exposed to N_2^{15} for two hours, fix nitrogen at a rate from 1 to 50 per cent of that of whole cells (Jose and Wilson, 1959), but there were many intact cells present. Azotobacter cells broken by ultrasonic vibration give good N_2^{15} fixation, too, the enzyme being in the participate fraction (Nicholas and Wilson, 1962). The best success has been with *Cl. pasteurianum*, similarly broken up by sonic vibration (Wilson and Burris, 1960; Carnahan *et al.*, 1960; Burris, 1962); this preparation in presence of pyruvate or (less effectively) ketoglutarate fixes about half as much N_2 as whole cells, and the enzyme is strictly soluble.

Spectroscopic observation of such preparations from Azotobacter and nodules indicated that both flavoprotein and cytochromes could be reduced by hydrogen and reoxidised by N_2 (Hamilton *et al.*, 1957), while Clostridium extracts show only the flavin being so reduced and reoxidised, as might be expected in an anaerobe. Hydrogen gas would normally be activated by the enzyme *hydrogenase*, and indeed both in Azotobacter and Rhodospirillum (but not in Clostridium) the hydrogenase content is higher when they are growing with No gas than when growing on ammonium or glutamate. But since the normal supply of hydrogen for reduction would be metabolic, i.e., H* + £, and would come via the usual oxidation-reduction mechanisms, the role of hydrogenase in fixation may have been made too much of. In any case the Clostridium enzyme has allowed a rigid test of the role of intermediates (Burris, 1962); if, after N_2^{15} fixation, unlabelled NH_2OH was added to the enzyme, it could be re-isolated as an oxime, and if any $N^{15}H_2OH$ were present, the oxime would be labelled. It was not.

Hydrazine similarly received no label whatever. In the same experiments ammonia became intensely labelled. Thus the N_2 is probably reduced directly or indirectly to ammonia on the enzyme. The dominating role of a reduction process fits well with the fact that within nodules the rhizobia must be under anaerobic conditions.

The first step must be for the N_2 to be bound. Fixation in Azotobacter is inhibited specifically and competitively by N_2O and by azide, N_2-NH (Roberts, 1959). In both these, the distance between the N atoms is nearly the same as in nitrogen gas, which could indicate that the enzyme binds both N atoms, and then its action is fundamentally to break the bonds between them. The bond energy of Ni-N is 225,000 calories, so that very large energy is needed to separate them. Because in cytochrome oxidase the atoms of O_2 are separated through a change in valency of the metal (iron) in the enzyme, so one visualises the same thing here, only with more extreme change of valency because of the higher energy requirement. Molybdenum, having valencies from 2 to 6, is thus most appropriate. Furthermore, the function of the nitrogenase is not only to bind the N_2, but also to bring it into contact with protons; then, presumably by the change of metallic valency, to donate enough electrons to allow the hydrogen to combine, *as* shown below, E being the enzyme. Finally the two amino groups would be removed by keto acids, while the electron-transfer chain reduces the metal back to its original valency state. But understanding of the details is for the future.

Bibliography

Ahuja, Sonia.: *Plant Diseases,* Vishvabharti, New Delhi, 2005.

Alfred, Steferud: *Diseases of Vegetable Crops,* Biotech Books, Delhi, 2005.

Amarjit, S. Basra: *Plant Growth Regulators in Agriculture and Horticulture,* International Book, Lucknow, 2004.

Arvind, M. Dhopte and Manuel Livera M.: *Principles and Techniques for Plant Scientist,* Agrobios, Jodhpur, 2002.

Babcock, E.B.: *Genetics and Plant Breeding,* Agrobios, Jodhpur, 2001.

Banerjee, Utpal: *Mineral Requirement and Growth in Plants,* Mangal Deep Publication, Jaipur, 2006.

Basra, Amarjit S.: *Plant Growth Regulators in Agriculture and Horticulture: Their Role and Commercial Uses,* International Book, Lucknow, 2004.

Beale, Helen Purdy: *Bibliography of Plant Viruses,* Columbia University Press, New York, 1976.

Bhattacharya, V.C.: *Phytotaxonomy: Journal of Association for Plant Taxonomy,* Deep Publications, New Delhi, 2002.

Bradbury, J.F.: *Guide to Plant Pathogenic Bacteria,* C. A. B. International, England, 1986.

Bridge, Paul Peter Jeffries, David R. Morse and Peter R. Scott.: *Information Technology, Plant Pathology and Biodiversity,* New India Publication, New Delhi, 2005.

Browne, F.G.: *Pests and Diseases of Forest Plantation Trees*, Clarendon Press, England, 1968.

Bukhari, A.Z.: *Encyclopaedia of Plant Ecology*, Anmol, New Delhi, 2007.

Chakraborty, Supriyo: *Plant Molecular Genetics*, Scientific, Jodhpur, 2005.

Chakravarti, B.P.: *Methods of Bacterial Plant Pathology*, Agrotech, Udaipur, 2005.

Chandra, Pravin Trivedi.: *Nematode Management in Plants*, Scientific, Jodhpur, 2003.

Chandra, Pravin Trivedi: *Plant Protection: A Biocontrol Approach*, Aavishkar, Jaipur, 2003.

Chandra, Trivedi Pravin: *Plant Pest Management*, Aavishkar, Jaipur, 2002.

Chattopadhyay, S.B.: *Manual of Plant Pathological Techniques*, Naya Udyog, Kolkata, 2003.

Chaube, H.S. and Ramji Singh: *Introductory Plant Pathology*, IBDCO, Lucknow, 2001.

Das, N.C.: *Ferns and Fern-Allies of Tripura*, International Book Distributors, Dehradun, 2007.

Desai, Lakshman: *Molecular Plant Pathology*, Paragon International Publication, New Delhi, 2007.

Dhankhar, O.P.: *Research Methods in Plant Sciences: Allelopathy*, Scientific Publication, Jodhpur, 2007.

Dronapushpi, R. B., Lamiaceae V. Singh: *Monograph on Indian Leucas*, Scientific Publishers, Jodhpur, 2001.

Felt, E.P.: *Plant Galls and Gall Makers*, Agrobios, Jodhpur, 2001.

Gautam, N.C.: *Plant Biotechnology*, Shree Publication, New Delhi, 2006.

Ghosh, S.K.: *A Glimpse on Terminology of Agricultural Entomology*, International, Lucknow, 2001.

Gopal, B. Singh and Aruna Kumari: *Terminology on Plant Physiology*, B.S. Publication, Hyderabad, 2005.

Goswami, C.L. and Politycka B.: *Research Methods in Plant Sciences,* Scientific Publication, Jodhpur, 2007.

Grierson, Don: *Plant Genetic Engineering,* Springer, New Delhi, 2007.

Gupta, G.P.: *Plant Cell Biology,* Discovery, New Delhi, 2004.

Gupta, G.P.: *Plant Pathology,* Discovery, New Delhi, 2004.

Gupta, P.K: *Genetics and Biotechnology in Crop Improvement,* Rastogi, Meerut, 1998.

Gupta, Rajni and Mukerji K.G.: *Microbial Technology,* APH, Delhi, 2001.

Gupta, S.K.: *Plant Breeding: Theory and Techniques,* Agrobios, Jodhpur, 2000.

Harmeet, Kour: *Encyclopaedia of Plant Physiology,* Anmol, New Delhi, 2006.

Hayes, H.K. and Garber R. J.: *Breeding Crop Plants,* Asiatic, Delhi, 2002.

Hemantaranjan, A.: *Plant Physiology and Biochemistry Plant Molecular Biology,* Scientific, Jodhpur, 2000.

Herman, S.: *Paxton's Botanical Dictionary,* IBD, Dehradun,1980.

Holliday, Paul.: *A Dictionary of Plant Pathology,* Cambridge University Press, Cambridge, 1998.

Howes, F.N.: *Vegetable Gums and Resins,* Scientific, Jodhpur, 2001.

Ignacimuthu, S. and Alok Sen.: *Strategies in Integrated Pest Management : Current Trends and Future Prospects,* Phoenix, New Delhi, 2002.

Irfan, A. Khan and Atiya Khanum: *Role of Biotechnology in Medicinal and Aromatic Plants,* Ukaaz, Hyderabad, 2002.

Jain, S.K.: *Index to Plant and Animal Names,* Deep Publications, Delhi, 2001.

Jha, D.K.: *Laboratory Manual on Plant Pathology,* Pointer, Jaipur, 2004.

Jha,Timir Baran and Biswajit Ghosh.: *Plant Tissue Culture : Basic and Applied*, Universities Press, Hyderabad, 2005.

Johri, R.M. Sneh Lata and Kavita Tyagi: *A Textbook of Bryophyta*, Dominant, New Delhi, 2004.

Kapoor, B.B.S. and Khatri, N.K.: *Management of Plant Diseases*, Madhu Publication, Bikaner, 2004.

Kapoor, S.L.: *Phytotaxonomy : Journal of Association for Plant Taxonomy*, Deep Publication, New Delhi,2006.

Karanth, Basavanna: *Selected Readings in Plant Genetics and Biotechnology*, Book Enclave, Jaipur, 2001.

Kendall, R. Lamkey and Michael Lee.: *Plant Breeding*, Blackwell, New Delhi, 2006.

Khan, Aslam: *Plant Anatomy and Physiology*, Kalpaz, Delhi, 2001.

Kirtikar, K.R.: *The Poisonous Plants of Bombay*, Scientific, Jodhpur, 2003.

Kumar, Dipak Mishra and Singh N.P.: *Endemic and Threatened Flowering Plants of Maharashtra*, Botanical Survey of India, Calcutta, 2001.

Kumar, U. and Sharma A.K.: *Plant Biotechnology and Biodiversity Conservation*, Agrobios, Jodhpur, 2001.

Kumar, U.: *Methods in Plant Tissue Culture*, Agrobios, Jodhpur, 2003.

Kyte, Lydiane and Kleyn John: *Plants From Test Tubes : An Introduction to Micropropagation*, Panima Publication, New Delhi, 2006.

Lamiaceae, R. Br. and Singh. V.: *Monograph on Indian Leucas*, Scientific Publishers, Jodhpur, 2001.

Macmillan, H.F.: *Handbook of Tropical Plants*, Anmol Publication, New Delhi, 2001.

Maheshwari, D.K. and Dubey R.C.: *Innovative Approaches in Microbiology*, Bishen Singh Mahendra Pal Singh, Dehradun, 2001.

Maheshwari, J.K.: *Recent Researches in Plant Anatomy and Morphology*, Scientific, Jodhpur, 2001.

Maloy, Otis C. and Timothy D. Murray.: *Encyclopedia of Plant Pathology*, J. Wiley, New York, 2001.

Mazumdar, B.C.: *Photoperiodism and Vernalization in Plants*, Daya, Delhi, 2005.

Meena Francis.: *Biotech's Dictionary of Botany*, Biotech, Delhi, 2005.

Miller, Edwin Cyrus: *Plant Physiology: With Reference to the Green Plant*, Biotech Books, Delhi, 2005.

Mishra, Shubhrata R.: *Mollicutes and Plant Diseases*, Discovery Publishing House, New Delhi, 2004.

Mishra, Shubhrata R.: *Photosynthesis in Plants*, Discovery Publication, New Delhi, 2004.

Mukerji, K.G. and Sharma A.K.: *Glimpses in Botany*, APH, Delhi, 2000.

Mukhopadhyay, N.C.: *Plant Taxonomy*, Aavishkar Publication, Jaipur, 2006.

Naidu, C.K.: *Plant Growth and Development*, Pointer, Jaipur, 2004.

Nehra, Sampat: *Plant Diseases : Biocontrol Management*, Aavishkar, Jaipur, 2001.

Nyvall, Robert F.: *Field Crop Diseases Handbook*, Iowa State University Press, Ames, 1999

Pal, Vijai.: *Research Methods in Plant Sciences: Allelopathy: Plant Pathogens*, Scientific, Jodhpur,2004.

Panda, S.K.: *Advances in Stress Physiology of Plants*, Scientific, Jodhpur, 2002.

Pandey, Ashok Carlos R. Soccol, Jose A. Rodriguez-Leon and Poonam Nigam: *Solid-State Fermentation in Biotechnology: Fundamentals and Applications*, Asiatech Publishers, New Delhi, 2001.

Pareek, L.K.: *Trends in Plant Tissue Culture and Biotechnology*, Agrobios Jodhpur, 2005.

Paul, Bridge Peter Jeffries and David R. Morse: *Information Technology, Plant Pathology and Biodiversity*, New India Publication, New Delhi, 2005.

Prakash, Ram Srivastava: *Neem and Pest Management*, IBDCO, Lucknow, 2001.

Pratap, Reddy N. and Pullaiah T.: *Gleanings in Plant Sciences*, Regency, New Delhi, 2004.

Pravin, Chandra Trivedi: *Glimpses in Plant Sciences and Biotechnology*, Pointer, Jaipur, 2003.

Pravin, Chandra Trivedi: *Plant Pathology*, Pointer, Jaipur, 2001.

Purcell, A. H.: *Advances in the Understanding of Pierce's Disease and its Insect Vectors*, University California, Davis, 1982.

Purohit, S.S. and Arvind Kumar: *Plant Physiology: Fundamentals and Applications*, Agrobios, Jodhpur, 2001.

Purohit, S.S. and R. Ranjan: *Plant Water Absorption and Transpiration*, Agrobios, Jodhpur, 2003.

Purohit, S.S., Bohra, S.P. and Bora K.K. : *Agro's Dictionary of Plant Physiology and Biochemistry*, Agrobios, Jodhpur, 2001.

Purohit, S.S.: *Plant Tissue Culture*, Student Edition, Jodhpur, 2004.

Rabindra, Narain and Surendra Naha: *Genetic Engineering in Plants*, Dominant Publication, New Delhi, 2006.

Rajak, R.C.: *Biotechnology of Microbes and Sustainable Utilization*, Scientific, Jodhpur, 2002.

Rajan, S. Sundara: *Plant Morphology*, Anmol, New Delhi, 2002.

Rajni, Sharma: *An Introduction to Plant Physiology*, Campus Books, New Delhi, 2004.

Ram, Reddy S. and Reddy S.M.: *Microbial Physiology*, Scientific, Jodhpur, 2005.

Ramachandra, P. Reddy and H. Ramakrishna: *Gleanings in Botanical Research: Current Scenario*, Dattsons, Nagpur, 2005.

Rana, J.C. and Sharma B.D.: *Plant Genetic Resources of Western Himalayas : Status and Prospects*, Bishen Singh Mahendra Pal Singh, Dehradun, 2005.

Ranjan, R., Purohit S.S. and Prasad V.: *Plant Hormones: Action and Application*, Agrobios, Jodhpur, 2003.

Rao, M. Lokeswara: *Plant Biodiversity of World*, International, Dehradun, 2004.

Raymond, Adams Dutcher, Clifford O. Jensen and Paul M. Althouse: *Introduction to Agricultural Biochemistry*, Biotech Books, Delhi, 2005.

Sagwal, S.S.: *Dictionary of Botany*, Dehra Dun, 2003.

Saraswat, V.N., Bhan V.M. and Yaduraju N.T.: *Weed Management*, Indian Council of Agricultural Research, New Delhi, 2003.

Schopfer, W.H. and Norbert L. Noecker: *Plants and Vitamins*, Asiatic Publication, Delhi, 2005.

Sharma, Bhuvan: *Plant Biotechnology*, Paragon International Publication, New Delhi, 2007.

Sharma, Rajni: *An Introduction to Plant Anatomy*, Campus, New Delhi, 2004.

Sharma, Ramphal: *Plant Biochemistry*, RBSA Publication, Jaipur, 2006.

Shubhrata, R. Mishra.: *Bacterial Plant Diseases*, Discovery, New Delhi, 2003.

Shubhrata, R. Mishra: *Morphology of Plants*, Discovery Publishing House, New Delhi, 2004.

Shubhrata, R. Mishra: *Mollicutes and Plant Diseases*, Discovery Publishing House, New Delhi, 2004.

Singh, M.P. and Chaudhary S.B.: *A Textbook of Practical Botany*, Daya, Delhi, 2005.

Singh, N.P. and Sanjappa M.: *Fascicles of Flora of India*, Botanical Survey of India, Kolkata, 2006.

Singh, S.K.: *Plant Ecology*, Campus Books, New Delhi, 2005.

Srivastava, R.C.: *Plant Resources of Uttar Pradesh*, Oriental Enterprises, Dehradun, 2006.

Subramanian, D.: *Monograph on Indian Charophyta*, Bishen Singh Mahendra Pal Singh, Dehradun, 2002.

Sundara, S. Rajan: *Physiology of Transport in Plants*, Anmol, New Delhi, 2003.

Susheela, M. Das.: *Latest Portfolio of Theory and Practice in Plant Anatomy*, Dominant Publication, New Delhi, 2003.

Thind, T.S.: *Diseases of Field Crops and their Management*, Daya, Delhi, 2005.

Trivedi, P.C.: *Plant Biotechnology: Perspectives and Prospects: Festschrift in Honour of Prof. C.P. Malik*, Pointer, Jaipur, 2006.

Trivedi, P.C: *Plant Biotechnology:Recent Advances*, Panima, New Delhi, 2000.

Tyagi, I.D.: *Plant Breeding and Genetics at a Glance*, South Asian, New Delhi, 2005.

Uniyal, B.P: *Plant Records*, Bishen Singh Mahendra Pal Singh, Dehradun, 2004.

Utpal, Banerjee: *Mineral Requirement and Growth in Plants*, Mangal Deep Publication, Jaipur, 2006.

Venkata, R.R., Raju, C. Sudhakar, B. Raviprasad Rao and S. Thimma Naik.: *Recent Trends in Plant Sciences*, Regency, New Delhi, 2005.

Verma, V.: *Textbook of Plant Physiology*, Ane Books India, New Delhi, 2007.

Vyas, S.P. and Kohli D.V.: *Methods in Biotechnology and Bioengineering: For Pharmacy and Other Biosciences*, C.B.S., New Delhi, 2002.

Walia, R.K. and Ram Singh: *Research Methods in Plant Sciences: Allelopathy: Plant Protection*, Scientific, Jodhpur, 2004.

Welch, Ross M. and Kalyan Singh: *Perspectives on the Micronutrient Nutrition of Crops*, Scientific, Jodhpur, 2001.

Zamir, K. Punja: *Biochemistry, Molecular Biology and Genetic Engineering*, International, Lucknow, 2005.

Index

A

B

C

D

E

O

P

Q

❑❑❑